X-RAY FLUORESCENCE ANALYSIS OF ENVIRONMENTAL SAMPLES

edited by

Thomas G. Dzubay

Research Physicist
Environmental Science Research Laboratory
U.S. Environmental Protection Agency
Research Triangle Park, North Carolina

ANN ARBOR SCIENCE
PUBLISHERS INC
P.O. BOX 1425 • ANN ARBOR, MICH. 48106

EDITORIAL BOARD

David C. Camp
University of California
Lawrence Livermore Laboratory
Livermore, California

John R. Rhodes
Columbia Scientific Industries
Austin, Texas

Joseph M. Jaklevic
University of California
Lawrence Berkeley Laboratory
Berkeley, California

Copyright © 1977 by Ann Arbor Science Publishers, Inc.
230 Collingwood, P. O. Box 1425, Ann Arbor, Michigan 48106

Library of Congress Catalog Card No. 76-22238
ISBN   0-250-40134-7

Manufactured in the United States of America
All Rights Reserved

This book was edited by Thomas G. Dzubay in his private capacity. No official
endorsement by the U.S. Environmental Protection Agency or any other agency
of the federal government is intended or should be inferred.

FOREWORD

To safeguard the quality of the air we breathe it is necessary to continue the development of improved analytical methods for identification, characterization and measurement of pollutants at emission sources and in the ambient atmosphere. In addition to their use in routine monitoring, such improved methods are needed for laboratory research, field investigations, and development of models to describe the transport and transformation of pollutants from emission sources to the ambient atmosphere. These activities necessitate that we have available the best possible means to collect and analyze organic and inorganic constituents in vapors and in aerosols. The process of seeking out and developing new analytical approaches requires that we evaluate the feasibility of application of a wide range of techniques to environmental problems.

X-Ray fluorescence analysis has proved to be one of the more successful of these techniques. Broadly speaking, our interest in X-ray fluorescence analysis and other techniques can be divided into two aspects. The first of these involves the careful and deliberate development of collection and sampling capabilities along with fully adequate calibration procedures. The second aspect is concerned with application of X-ray fluorescence analysis to those environmental problems for which it is appropriate. This book emphasizes the first of these aspects, but it is important that we apply techniques for evaluation of environmental problems as soon as they are adequately developed.

There are several important applications of X-ray fluorescence analysis to problems in the atmospheric science. One is the evaluation of the abundance of various elements in the fine and coarse fractions of airborne particulate matter. A second is investigating the geographic distribution and time trends of elements in aerosols. A third application involves the use of such elements as lead, sulfur, and vanadium to serve as the tracers for the long-range transport of pollutants. A fourth involves analyzing for catalytic substances such as manganese, vanadium, and iron from sources or urban plumes to relate composition to such phenomena as sulfate formation. A fifth involves relating emission of pollutants at sources to patterns of elemental concentrations in the ambient atmosphere. The importance of the latter cannot be overemphasized since a sound control strategy must be based on a well-established relationship between the emission of pollutants and their concentration after being dispersed into the atmosphere. It is no small task to determine such relationships because of the complex interactions and transformations that can occur. X-Ray fluorescence analysis can certainly provide highly useful information in all of these areas of interest.

A. P. Altshuller, Director
Environmental Sciences Research
Laboratory
U.S. Environmental Protection Agency

PREFACE

In recent years X-ray fluorescence spectroscopy has come into extensive use for monitoring the concentrations of trace elements in the environment. This has been brought about by several developments in the fabrication of modern X-ray fluorescence spectrometers, which have the potential to rapidly analyze trace levels of toxic elements in large numbers of aerosol and water pollution samples. In order to realize this potential, the scientific community has been challenged to develop reliable calibration standards and analysis procedures.

This volume brings together the experience of the scientists who have made significant contributions to the development of techniques for quantitative analysis of environmental samples using X-ray fluorescence. The chapters are organized into six sections, which cover the major problems encountered by the analyst.

The major hardware configurations for X-ray analysis of particulate samples are described in Section I. These include energy dispersive and wavelength dispersive analyzers, and X-ray and charged particle types of excitation systems. A detailed comparison of the elemental detection capabilities of three representative types of X-ray fluorescence analyzers is provided.

Section II details the methods for aerosol sampling. The use of dichotomous virtual impactors and adhesive-coated cascade impactors is described for the collection of size-fractionated aerosol particles on surfaces that are suitable for X-ray analysis. A thorough discussion of the properties of membrane filters for aerosol sampling is provided.

Methods for preparation of water samples prior to X-ray fluorescence analysis are described in Section III. This includes discussion of a very effective evaporation technique and a technique for chemically depositing ions on a surface suitable for X-ray analysis.

Section IV deals with calibration of X-ray spectrometers. A variety of methods using thin films, solution deposits, and particle deposits is presented.

Section V covers the problem of attenuation of X-rays in samples and presents a variety of procedures for calculating and measuring the attenuation correction factors for fine and coarse particles collected on membrane filters.

The problems of spectral analysis, treated in Section VI, include corrections for background, interelement interferences, gain and baseline shift, and X-ray attenuation. Discussions of linear least squares and nonlinear analysis procedures as well as novel approaches to background subtraction are given. Methods for reporting accuracy and detection limits on environmental samples are also described.

This volume should provide the reader with a fairly complete insight into approaches for performing quantitative analysis of air and water samples using X-ray fluorescence spectroscopy. It is hoped that such analyses will add to the knowledge required to improve the quality of our environment.

T. G. Dzubay

CONTENTS

**SECTION VI: MATHEMATICAL METHODS FOR ANALYSIS
OF X-RAY SPECTRA**

SECTION I

APPROACHES TO X-RAY ANALYSIS

PHOTON-INDUCED X-RAY FLUORESCENCE ANALYSIS
USING ENERGY-DISPERSIVE DETECTOR
AND DICHOTOMOUS SAMPLER

J. M. Jaklevic, B. W. Loo and F. S. Goulding

Lawrence Berkeley Laboratory
University of California
Berkeley, California

INTRODUCTION

In recent years X-ray fluorescence analysis has emerged as a very powerful technique for the elemental analysis of environmental samples. Several laboratories are currently applying the technique to the analysis of airborne particulate matter. The samples typically consist of uniform deposits of particulate matter collected on a thin clean substrate, making them ideally suited for nondestructive X-ray analysis.

Traditionally X-ray tubes have been employed to excite the fluorescence in samples, and crystal spectrometers were used to disperse and analyze the characteristic X-rays according to their wavelength. The development of semiconductor detectors has made possible the use of energy-dispersive spectrometers together with either heavy charged particle or photon excitation. The charged particles are typically 3 MeV protons or 16 MeV alpha particles that have been generated by an accelerator. Alternatively, photon excitation by monochromatic or broadband X-rays generated either in an X-ray tube or radioisotope source can be used.

Energy-dispersive analysis with charged particle excitation and wavelength-dispersive analysis are discussed in Chapters 2, 3 and 4 of this book. We shall describe an energy-dispersive X-ray fluorescence system using photon excitation, which has been designed for a large-scale aerosol sampling network. In addition to providing the required sensitivity for air particulate analysis, the X-ray unit has been designed for fully automatic operation as an integral part of a complete sampling, analysis, and data handling system. An automatic air sampler has been developed that is capable of acquiring particle size-segregated aerosol specimens that are compatible with the X-ray fluorescent unit. This discussion of the complete system is limited by the scope of the present chapter; previous publications give a more complete description of the techniques employed.[1-5]

X-RAY FLUORESCENCE SYSTEM BASIC CONCEPTS

A simplified schematic of an X-ray fluorescence technique is shown in Figure 1.1. Primary radiation is incident on the sample where it interacts to produce vacancies in the inner atomic shells, which then deexcite to produce the characteristic X-rays of interest. These X-rays from the sample are then detected, and their energies are measured by the semiconductor detector spectrometer.

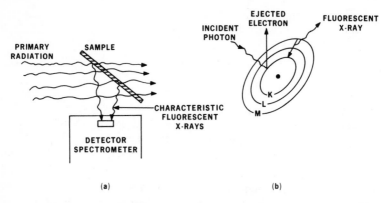

Figure 1.1 Schematic of the X-ray fluorescence technique.

The semiconductor detector spectrometer converts the energy of the incident X-ray into a voltage pulse whose amplitude is proportional to that energy. A multichannel analyzer is used to accumulate a histogram of the pulse amplitude spectrum. The energy resolution of the semiconductor detector spectrometer is more than adequate to separate X-ray lines from elements of adjacent atomic numbers. As such, it is capable of performing simultaneous multiple element analysis for typical aerosol samples. The area of the individual characteristic X-ray peaks in the spectrum is proportional to the concentration of the various elements in the sample.

DESIGN OF EXCITATION SYSTEMS

An X-ray tube was chosen as the primary excitation source because of its higher output compared to generally available radioisotope sources. The requirement to analyze 100 or more samples per day requires the highest possible counting rate. X-ray photons incident on the sample interact either by the photoelectric effect to produce the desired inner-shell atomic vacancies in the elements of interest or by scattering mainly from the atoms in the low atomic number substrate. These scattered X-rays constitute an unwanted background that sets the detection limit for the fluorescence measurement.

Figure 1.2 is a schematic of a spectrum produced by monoenergetic X-rays exciting a low atomic number filter substrate containing a few trace elements. The coherent (Rayleigh) and incoherent (Compton)

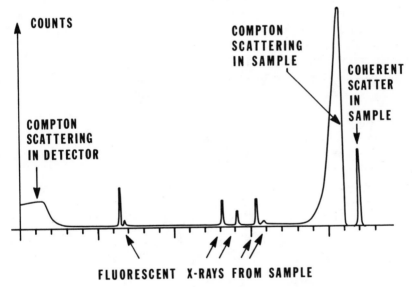

Figure 1.2 The response of the spectrometer for monoenergetic photons striking the sample. The matrix is assumed to be composed of light elements.

scattering peaks occur at or just below the energy of the incident X-rays. For good sensitivity the fluorescent X-rays of the elements must not overlap the scatter peaks. This effect favors the use of monoenergetic X-ray excitation sources rather than broadband excitation, which would distribute the scattered radiation over the entire range of energies.

However, the probability of producing fluorescent excitation of an element is greatest when the exciting X-ray energy just exceeds the binding energy of the electrons in the appropriate shells and falls off rapidly with increasing excitation energy. This implies that very low energy characteristic X-rays are not efficiently produced by monochromatic radiation of high energy, thus limiting the range of elements that can be sensitively measured with a single exciting energy. This effect compensates for some of the disadvantages of continuous excitation cited earlier. In some cases, such as the analysis for very light elements (Z < 20), the use of continuum excitation may give better sensitivity than monochromatic excitation.

To cover a broad range of elements with optimum excitation we have decided to use three measurements on each sample, using a different X-ray excitation energy for each. This is accomplished by operating in a secondary fluorescence mode in which the continuum radiation from the X-ray tube anode is incident on an external secondary target whose characteristic X-rays are then used to irradiate the sample. This method of generating nearly monoenergetic X-ray excitation has the advantage that it can be performed by mechanically switching the secondary target.

Figure 1.3 shows a cross section of the fluorescence geometry. The secondary target is in the form of a sheet metal enclosure that is moved

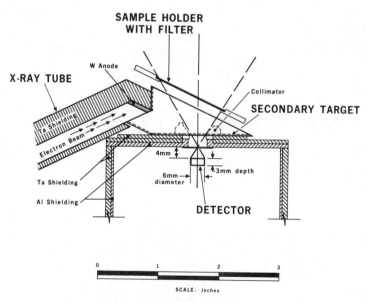

Figure 1.3 Sample-fluorescer-detector geometry employed in the secondary fluorescence mode of energy-dispersive analysis.

as an integral assembly with the collimators. The secondary targets, principle excitation energies, and range of atomic numbers analyzed by each are as follows:

(1) Ti, 4.5 keV, $13 \leqslant Z \leqslant 20$

(2) Mo, 17.4 keV, $20 \leqslant Z \leqslant 38$ together with the L X-rays for heavy elements Pb and Hg

(3) Sm, 40 keV, $38 \leqslant Z \leqslant 56$.

The X-ray tube is of a special design that projects the electron beam through a narrow tube onto a grounded anode. This allows for a tight geometry between secondary targets and the sample. Using this close coupled geometry we achieve maximum counting rates (15,000 counts/sec) with an X-ray tube power dissipation well below 100 watts.

DETECTOR CONSIDERATIONS

A low-background guard-ring detector is used in the spectrometer to ensure low background in the region below the scatter peaks.[6] A pulsed-light feedback electronic system is used to provide good energy resolution and high counting rate capabilities. Of particular interest in our most recent X-ray fluorescence system is the use of a pulsed X-ray tube to provide a greatly increased output counting rate. In this system an X-ray tube control grid is operated in a feedback loop with the detector output in such a way that the tube is pulsed off when the electronics is processing a signal and turned on only when the system is prepared to analyze an event.[7] The effective output counting rate can be increased two to three times using this method while still limiting pulse pile-up to a tolerable value.

SAMPLE FORM AND CALIBRATION

The present system has been designed to accommodate air-particulate samples collected by filtration. These samples are thin, uniform deposits of particles collected on clean membrane filters consisting of cellulose, polycarbonate, or other hydrocarbons. Glass fiber filters generally contain substantial amounts of elemental contaminates that interfere with the analysis and are not suitable for these applications.

The calibration of the X-ray analyzer consists of converting the observed characteristic X-ray counting rates to concentration of elements expressed as ng/cm^2 on the filter. A simple model can be used to describe the relative efficiency for detecting X-rays as a function of concentration of a given element. For samples in which the attenuation of the fluorescence X-rays can be neglected, the counting rate for a given characteristic X-ray can be expressed as

$$I_i = I_o \, G \, K_i \, \rho_i$$

where I_o is the X-ray source intensity, G is a geometry and efficiency factor, ρ_i is the concentration of element i, and K_i is the relative X-ray excitation cross section. In this simple case and when using monoenergetic excitation, K_i can be calculated from a straightforward physical model of X-ray photoelectron interactions (see Chapter 12).[8]

In cases where the X-ray energies are low enough that absorption either within the filter matrix or from the individual particles becomes important, an added attenuation correction must be included. It is the magnitude of this correction and the ability to estimate it accurately which constitutes the major limitation of the XRF technique in the case of light elements. Several papers discussing both this problem and calibration techniques in general are included in this volume.

Figure 1.4 illustrates the operation of the X-ray fluorescence system in a completely automatic mode for large-scale analysis applications. The aerosol samples are collected on membrane filters mounted in a standard 5-cm x 5-cm holder, which is compatible with the standard 35-mm slide

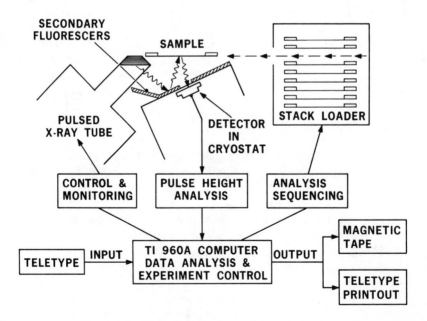

Figure 1.4 Block diagram of complete analysis system.

projector hardware used in the automatic sample handling. An automatic stack loader is used to sequence the samples through the analysis system. The operation of the stack loader, secondary target manipulator, and X-ray tube controller is sequenced by a small computer, which is also used as a pulse height analyzer and data output controller. A spectrum analysis program is included in the computer software and the data are written on the output tape both in their original spectral form and as reduced concentrations per unit area.[1,5] The output data are subsequently corrected for filter attenuation, particle size and possible interelement effects before being stored in a data bank.[9,10]

AIR SAMPLER DESIGN

Routine collection of aerosol particles on a membrane-type filter is a straightforward problem requiring that a known volume of air be drawn through the filter using a vacuum pump. Flow rates of 5 to 10 liters/ min/cm^2 of the filter can easily be achieved. At such rates it is possible to observe major elemental constituents of the urban aerosol in samples collected in 30 minutes or less. More typically, sample collection intervals will be 2 to 6 hours or longer depending upon the nature of the monitoring problem.

Recent air sampling requirements have emphasized the collection of particulate samples in separate size fractions. In particular, it is important to collect the two major components of the bimodal urban aerosol[11] as separate samples. It has been established that the fine particle accumulation mode with a peak at about 0.3 μm mass median diameter consists primarily of the combustion products, while natural and mechanically generated particles are usually larger than several microns in size. A size cut at about 2 μm corresponding to the minimum in the bimodal volume or mass distribution not only separates the particles according to their origin, and hence chemical properties, but also renders the procedure for X-ray attenuation corrections more manageable (see Chapter 16).[9]

The most widely used sampling technique for acquiring size segregated particles involves the removal of the particles from the air stream by impaction on a plate. Figure 1.5 illustrates a typical impactor stage. The air is directed through a narrow slit onto an opposing surface causing the streamlines to turn abruptly. Particles in the air stream experience the opposing forces of the viscous interaction in the air and the inertial tendency to travel in a straight line. The viscous force is proportional to the cross-sectional area of the particle whereas the inertial "force" is proportional to the particle volume. If the air velocity and average turning radius are suitably adjusted, particles above a certain effective diameter

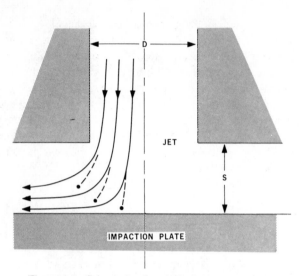

Figure 1.5   Schematic of a single impactor stage.

are caused to impact onto the plate while the smaller particles remain in the air stream. A number of commercially available air samplers utilize this process to collect samples in a sequence of successively smaller particle size ranges.

This method possesses several disadvantages with respect to subsequent X-ray fluorescence analysis. Patterns of particle depositions are images of the narrow slits used in the individual impactor stage. Such small area, high density deposits do not satisfy the criteria for uniformity and thickness established earlier. The conventional impactor also exhibits certain inherent disadvantages. Particles incident on the impaction surface often fail to adhere and become reentrained in the air flow. This particle bounce phenomenon results in large particles being collected in the smaller particle fraction. In the case where heavy deposits are accumulated on the impaction surface, the surface particles can be blown off by the air stream and recollected by a later stage.

To avoid these problems and to provide aerosol specimens suitable for X-ray fluorescence analysis, we have designed an air sampler based on the principle of virtual impaction.[4]   Instead of a physical plate for an impaction surface, the aerodynamics are designed so that the impaction occurs into a separate air volume that can then be drawn through a filter. Figure 1.6 illustrates a single circular jet that is impacted into a tube with a portion of the flow Q_1 drawn into the tube with a majority of the flow Q_2 continuing around the impaction tube. The large particles should remain in the flow Q_1, whereas the small particles will be divided

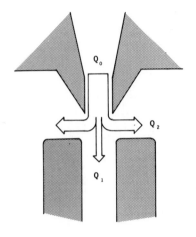

Figure 1.6   Schematic of a virtual impaction orifice.

between Q_1 and Q_2 in proportion to the relative air flows. It is obvious that such a design is free from the problem of particle bounce and blow-off.

Following an extensive program of measuring the properties of such an impactor stage, we have designed a two-stage virtual impactor with a single cut point at 2.4 μm particle diameter. This dichotomous air sampler, shown schematically in Figure 1.7, operates at an input flow rate of 50 liters/min. The outputs of the device are passed through a pair of membrane filters similar to those discussed earlier. To ensure stable operating conditions, the flow division Q_1/Q_2 and total flow Q_0 are maintained constant by monitoring the pressure drop across a fixed orifice in the system. This compensates for any change in flow caused by the increased loading on the membrane. Figure 1.8 shows the effectiveness of the size separation process. The fraction A/A+B is the fraction of particles that is collected as large particles. The fraction of particles lost through the impaction on various inner surfaces is also shown. The solid particle losses have been reduced to a tolerable level, which compares favorably with the performance of other types of air samplers.

An automatic air sampler incorporating the dichotomous virtual impactor as a first stage has been designed, and ten such units have been operating in St. Louis for the past year. The sampler accommodates the membrane filter holder in pairs that are automatically sequenced according to a predetermined sampling schedule. The 5-cm x 5-cm samples are contained in conventional 36-slide cartridges used in photographic slide projectors. These cartridges are changed as needed and returned to a central laboratory for analysis by X-ray fluorescence.

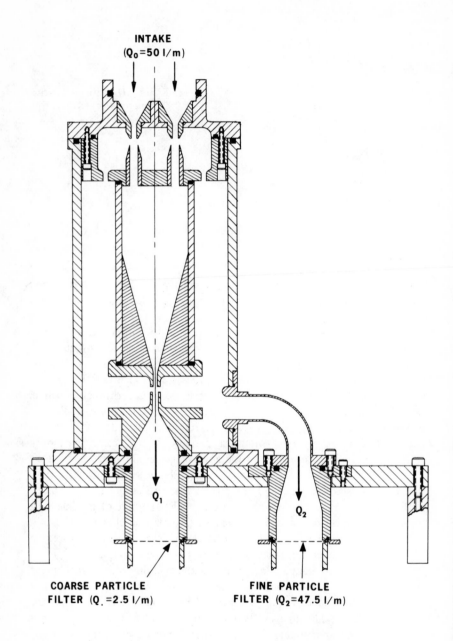

Figure 1.7   Cross-sectional drawing of a dichotomous virtual impactor. Filters A and B are membrane-type filters.

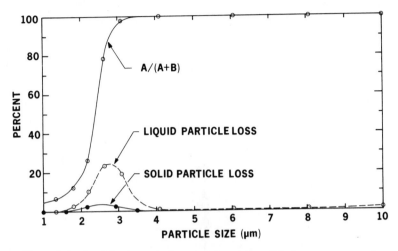

Figure 1.8   Particle size cut characteristics and particle loss measurements
for the dichotomous sampler.

Also included in the program are total mass measurements using a
beta-particle attenuation method.[3]   The filters are gauged before and after
exposure to the aerosol to determine the mass differences.   Such mass
differences can be determined to an overall accuracy of ± 10 $\mu g/cm^2$
with 30-second measurements.

RESULTS

Figure 1.9 illustrates X-ray fluorescence data taken on a pair of filters
that were exposed in the automatic dichotomous sampler.   The difference
in elemental distribution from the two particle-size ranges is apparent.
The elements lead and bromine are found predominantly in the fine particle fraction, whereas calcium and iron are contained mostly in the coarse
particles.   The fine aerosol particles (< 2 μm diameter) are assumed to
originate from combustion reactions such as automotive engines for lead
and bromine.   The coarse particles originate from mechanical processes
such as wind blown soil dust.[12,13]
Analysis of this type of data from a time sequence of samples exposed
at several sites yields information on the time and spatial variation of
elemental contaminants in the atmosphere.   During the past year over
15,000 such samples have been collected using automatic dichotomous
samplers at ten selected sites in the St. Louis area.   These are currently
being analyzed using photon-excited energy-dispersive analysis, and data

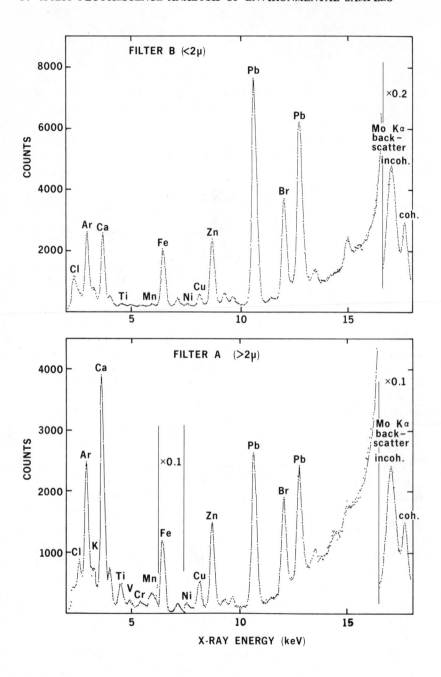

Figure 1.9   Photon excited energy-dispersive analysis spectra of a membrane filter containing the fine particle fraction (upper) and the coarse particle fraction (lower).

are being incorporated into the EPA Regional Air Pollution Study (RAPS) data bank.

Figure 1.10 shows plots of fine and coarse particle concentrations for the elements sulfur and lead in air samples taken over three days beginning on July 29, 1975. The fact that sulfur and lead appear predominantly in the fine particle fraction indicates that their origin is in combustion processes from fossil fuel power plants (sulfur) and automotive exhaust (lead). The difference in diurnal patterns for the data acquired at an urban sampling station (RAPS site 103) reflects the periodic behavior of automobile traffic patterns in the lead concentration as opposed to the slowly varying sulfur levels. Figure 1.11 is a plot of the concentration of lead and sulfur as a function of the site for 10 sites over a 40-km radius of downtown St. Louis. The behavior of the sulfur concentration indicates a widespread background level over the entire region whereas the lead concentration reflects the local traffic densities at the sampling sites. For example, RAPS sites 122 and 124 are in relatively remote rural areas approximately 40 km from St. Louis, whereas sites 105, 106, 112 and 120 are in the St. Louis urban area.

CONCLUSION

Operating experience in using the photon-excited energy-dispersive X-ray fluorescence analysis system has demonstrated the applicability of this technique to large-scale air-sampling networks. This experience has shown that it is possible to perform automatic sampling and analysis of aerosol particles at a sensitivity and accuracy more than adequate for most air pollution studies.

ACKNOWLEDGMENTS

Work for this chapter was supported in part by the Energy Research and Development Administration under Interagency Agreement with the Environmental Protection Agency.

The authors wish to express their appreciation to the other members of the group who have contributed to the design, construction and operation of the aerosol analysis system. These include R. Gatti, B. Jarrett, D. Landis, N. Madden, J. Meng and W. Searles.

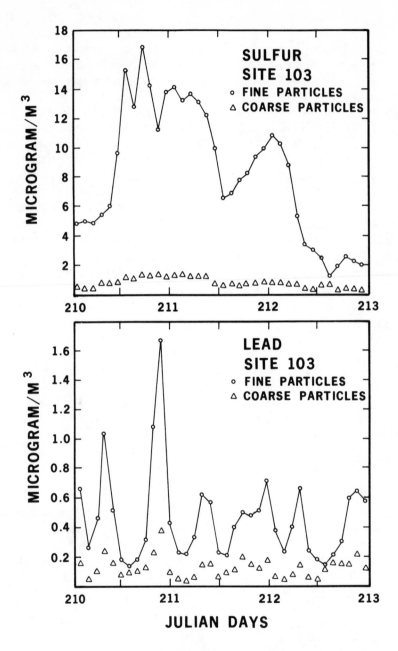

Figure 1.10   Sulfur and lead concentration as a function of time for an urban
sampling site.

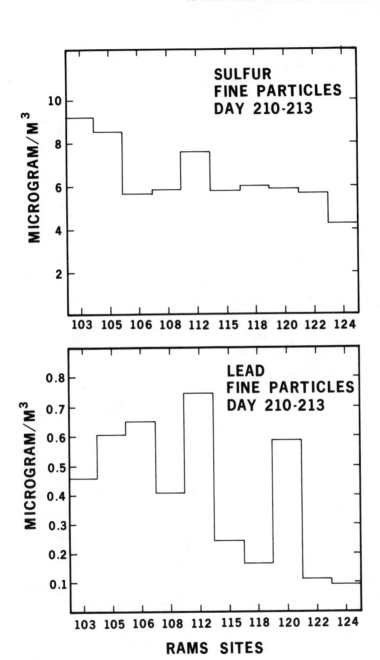

Figure 1.11 Sulfur and lead concentration for the complete network of stations averaged over a three-day sampling period in 1975.

REFERENCES

1. Goulding, F. S. and J. M. Jaklevic. "X-Ray Fluorescence Spectrometer for Airborne Particulate Monitoring," Environmental Protection Agency, Report No. EPA-R2-73-182 (1973).
2. Goulding, F. S. and J. M. Jaklevic. "Development of Air Particulate Monitoring Systems," Environmental Protection Agency, Report No. EPA-650-4-74-030 (1974).
3. Goulding, F. S., J. M. Jaklevic and B. W. Loo. "Fabrication of Monitoring System for Determining Mass and Composition of Aerosols as a Function of Time," Environmental Protection Agency, Report No. EPA-650/2-75-048 (1975).
4. Loo, B. W., J. M. Jaklevic and F. S. Goulding. "Dichotomous Virtual Impactors for Large Scale Monitoring of Airborne Particulate Matter," in *Fine Particles: Aerosol Generation, Measurement, Sampling and Analysis*, B. Y. H. Liu, Ed. (New York: Academic Press, 1976), pp. 311-350.
5. Jaklevic, J. M., F. S. Goulding, B. V. Jarrett and J. D. Meng. In *Analytical Methods Applied to Air Pollution Measurements*, R. K. Stevens and W. F. Herget, Eds. (Ann Arbor, Michigan: Ann Arbor Science Publishers, 1974), p. 123.
6. Goulding, F. S., J. M. Jaklevic, B. V. Jarrett and D. A. Landis. "Detector Background and Sensitivity of X-Ray Fluorescence Spectrometers," *Adv. X-Ray Anal.* 15:470 (1972).
7. Jaklevic, J. M., D. A. Landis and F. S. Goulding. "Energy Dispersive X-Ray Fluorescence Spectrometry Using Pulsed X-Ray Excitation," *Adv. X-Ray Anal.* 19:253 (1976).
8. Giauque, R. D., F. S. Goulding, J. M. Jaklevic and R. H. Pehl. "Trace Element Determination with Semiconductor Detector X-Ray Spectrometers," *Anal. Chem.* 45:671 (1973). See also Chapter 12 of this volume.
9. Dzubay, T. G. and R. O. Nelson. "Self-Absorption Corrections for X-Ray Fluorescence Analysis of Aerosols," *Adv. X-Ray Anal.* 18:619 (1975).
10. Loo, B. W., R. C. Gatti, B. Y. H. Liu, C. Kim and T. G. Dzubay. "Absorption Corrections for Submicron Sulfur Collected in Filters," Chapter 16 of this volume.
11. Whitby, K. T., R. B. Husar and B. Y. H. Liu. "The Aerosol Size Distribution of Los Angeles Smog," in *Aerosols and Atmospheric Chemistry*, G. M. Hidy, Ed. (New York: Academic Press, 1972), pp. 237-264.
12. Dzubay, T. G. and R. K. Stevens. *Envir. Sci. Technol.* 9:663 (1975).
13. Friedlander, S. K. *Envir. Sci. Technol.* 7:1115 (1973).

2

PROTON-INDUCED AEROSOL ANALYSES:
METHODS AND SAMPLERS

J. William Nelson

Department of Physics
Florida State University
Tallahassee, Florida

INTRODUCTION

Accelerator-produced beams of ions have been extensively used in research and industry. A recent survey[1] by I. L. Morgan of Columbia Scientific Industries Corp. lists 2500 accelerators in the U.S. with 80% of them in medical or industrial uses. Elemental quantitative analysis is perhaps the newest application. In this paper, two broad-range types of proton beam analyses and air particulate sampling devices designed for these methods will be described. Both proton-induced X-ray emission analysis (PIXE) and proton elastic scattering analysis (PESA) are employed to analyze aerosol samples using the Florida State University Tandem Van de Graaff Accelerator. These methods are fast, broad-range, absolute, subject to automation and well-suited for samples having areas of a few square millimeters and thicknesses of 1 mg/cm^2 (or approximately 10 μm). Samples of uniform density are most convenient; however, nonuniform samples may be analyzed by using a uniform proton beam.

AIR PARTICULATE SAMPLERS

To utilize to advantage this small sample size, two basically different types of air particulate samplers are employed. First, single orifice impactors adapted from a Batelle design[2] are used to gain particle size information and second, time sequence total filter samplers of Florida

State University (FSU) design ("streakers") are used to obtain a continuous time record.[3]   A top view of the latter sampler with its protective cover removed is shown as Figure 2.1.   A 2- x 5-mm air intake (upper right center) is driven to the left at a rate of 1 mm per hour by the synchronous motor.   A Nuclepore filter strip is placed over the intake to produce an 84-mm long strip sample of one week duration with two hour time resolution.   For operation, the structure holding the filter strip is rotated 180° from the pictured view so that air is admitted through the slot in the bottom plate, and an aluminum cover is provided to make the sampler weatherproof.   This sampler has been in field for two years in extremes of both winter and summer weather and has proven to be reliable.

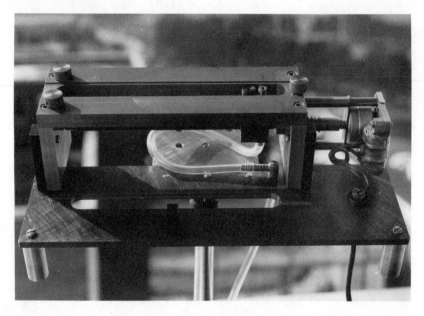

Figure 2.1   View of the continuous filter sampler. The 2- x 5-mm air intake (upper right center) is driven to the left at a rate of 1 mm per hour by the synchronous motor. A Nuclepore filter strip is placed over the intake to produce an 84-mm long strip sample in one week.

A photograph of the prototype of a newer circular model of this sampler is shown as Figure 2.2.   Considerable savings in size and simplicity are achieved by using circular geometry for the sample.   Additional features such as an optional impaction or prefilter stage are also included.   Thus, this new version will be capable of operation either as a total filter device or one with two size differentiations.

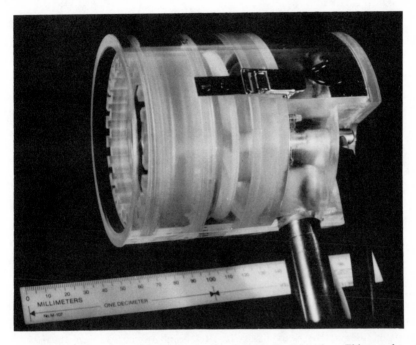

Figure 2.2   Side view of the prototype of a circular filter sampler. This sampler
replaces the larger linear sampler of Figure 2.1 and in addition has provisions for
an impaction or prefilter stage.

Typical samples from the single orifice impactor and linear "streaker"
are shown in Figure 2.3. The upper row contains particulate matter
samples impacted onto 100 $\mu g/cm^2$ polystyrene backings and a Nuclepore
afterfilter (upper left). For these samples of nonuniform areal density,
a uniform proton beam of 4.6 mm diameter is used in analysis. The
lower two samples are Nuclepore filter (0.4 μm pore size) strips exposed
in the streaker sampler. The center sample was obtained with continuous
motor drive of the orifice while the bottom sample shows discrete exposure of 2 x 5 mm areas. Both the impactor and streaker samplers operate
at flow rates of about 1 liter per minute. This modest pumping requirement permits the use of small pumps of various types (vane, diaphragm,
piston) for both indoor and outdoor application.

PROTON-INDUCED X-RAY EMISSION ANALYSIS

Beams of 10-50 nanoamperes permit proton-induced X-ray emission
analysis to be performed on these aerosol samples. Using an 80 mm^2
Si(Li) detector placed 60 mm from the sample, a spectrum containing

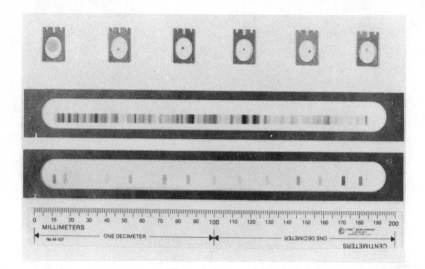

Figure 2.3   Typical air particulate samples suitable for proton beam analyses.

10 to 15 elements ranging from silicon to lead may be accumulated in a few minutes with a 4 to 5 MeV proton bombardment. In an equal time with the aid of a computer, a simultaneous least-squares fit to all the peaks and background can be performed.

In Figure 2.4, a streaker sample is shown being inserted into the sample holding chamber. After a plastic side is placed over this unit and the air pumped from the chamber, a proton beam of area 2 mm x 5 mm is incident upon the filter. Upon completion of a bombardment, a stepping motor advances the strip sample 2 mm for the next bombardment.

A typical spectrum obtained in this manner is shown in Figure 2.5. The line following the data points is the least squares fit while the line beneath the peaks represents the continuous background determined by the least squares fitting program. Considerable effort has been expended in developing this computer program, which obtains a fit to a spectrum using a measured library of peaks and uses nonlinear parameters in order that the energy calibration and peak resolution may be allowed to vary from one spectrum to the next. This program is described in detail by Kaufmann et al.[4]

For comparison with the filter sample spectrum of Figure 2.5, an impactor sample spectrum is presented as Figure 2.6. This sample consists of the larger particles (aerodynamic diameter > 4 μm) from 2.7 m³ of Tallahassee air collected at the first stage of a single orifice impactor on a paraffin-coated thin polystyrene backing. This linear plot with changes

Figure 2.4 A streaker sample being inserted into the sample holding chamber in preparation for proton bombardment.

of vertical scale shows clearly that the continuum background is minimized when thin backings of approximately 100 $\mu g/cm^2$ are used.

The overall detection efficiency versus atomic number of our present FSU system is shown in Figure 2.7. These values represent only those elements contained in our current computer program library, and thus not all elements are present. The values shown are not the usually quoted cross section but the readily usable number of photons per microcoulomb of protons for one microgram per square centimeter of element. The values were measured using samples whose areal densities were known from other means (primarily weighing).

PROTON ELASTIC SCATTERING ANALYSIS

Due to the degree of attenuation of softer X-rays by matter, X-ray analysis of the lightest elements is made difficult to impossible for practical sample loadings. Proton elastic scattering analysis (PESA) proves to

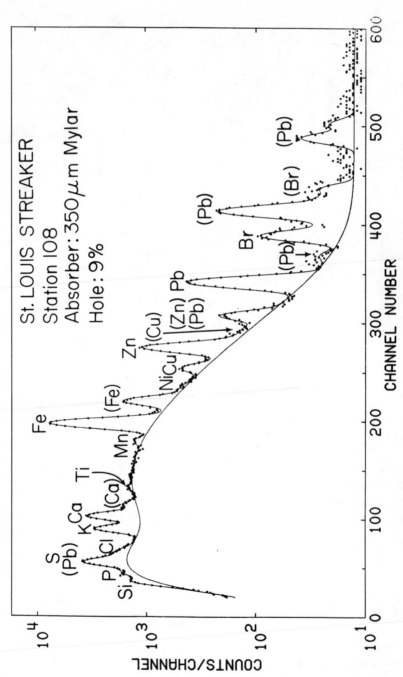

Figure 2.5   PIXE spectrum for a 2-mm wide region of a streaker sample.

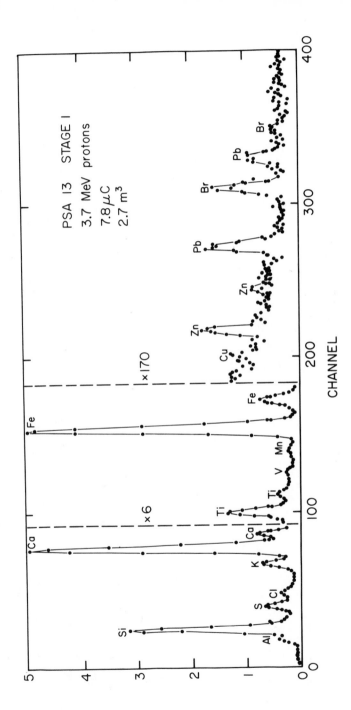

Figure 2.6 PIXE spectrum of a single orifice impactor sample.

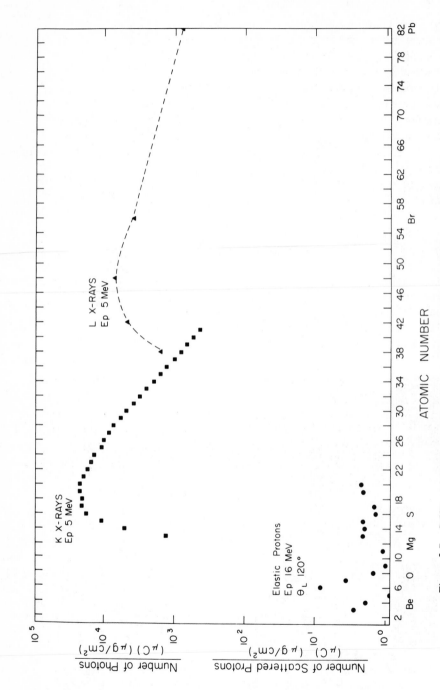

Figure 2.7 PIXE and PESA overall efficiency versus atomic number for the current FSU system.

be a useful complement to PIXE since it has almost identical sample requirements and is also nondestructive of the sample.[5]   With modern
commercially available solid state proton detectors and 16 MeV incident
protons it is possible to resolve the light elements up through chlorine.
Two other advantages to the use of 16 MeV protons are their smaller
specific energy loss (allowing the use of thicker samples) and relative
freedom from resonant behavior of the excitation function (allowing
quantitative results without the need for sample thickness corrections).

On the negative side, the nuclear scattering cross sections are approximately 1000 times smaller than the atomic cross sections for PIXE.   This
may, be seen for the FSU system in Figure 2.7 in which the PESA overall efficiency is displayed in the same terms as that for PIXE.   The effect
of this lower efficiency is greatly counterbalanced by the lower backgrounds
for PESA spectra, the use of larger beam currents and the usual presence
of greater amounts of many light elements in air particulate matter.

The FSU PESA system uses the same sample chamber as used for
PIXE.   A view of the solid state proton detector being held near the
chamber is shown in Figure 2.8.   Scattered protons emerge from the
vacuum chamber through a Mylar window and are collimated by an
elliptical slot into the large area (6 x 50 mm) detector.   A typical
spectrum for a Tallahassee impactor sample is shown in Figure 2.9.   All

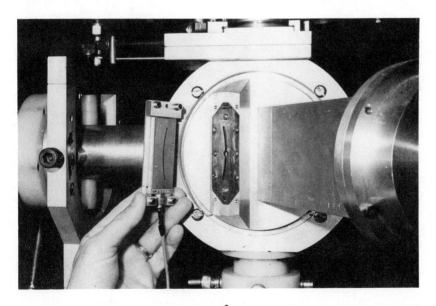

Figure 2.8   View of a large area (300 mm^2) solid state proton detector being held
near the Mylar window of the sample chamber.

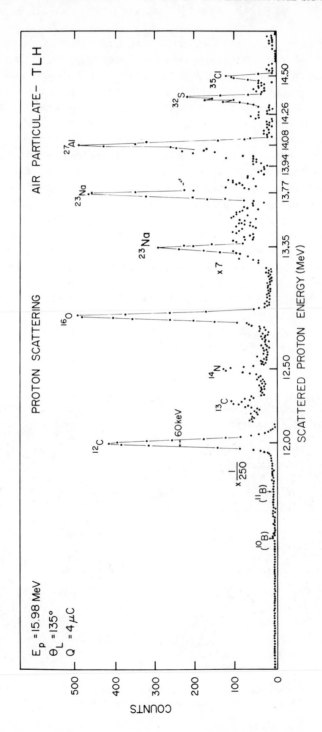

Figure 2.9 PESA spectrum for a Tallahassee impactor sample.

peaks shown are due to elastically scattered protons except the peak at
13.35 MeV, which is an inelastic scattering peak due to the first excited
state of sodium. A computer analysis program for PESA (similar to that
developed for PIXE) is expected to be completed in the near future.

APPLICATIONS

One application of these samplers and methods was in the sulfate
emission study of catalytic converter equipped automobiles at the General
Motors Proving Ground during October 1975. Towers on the upwind
and downwind sides of a four-lane road were extensively instrumented
and included FSU single orifice impactors and linear streaker samplers.
In the upper portion of Figure 2.10, an impactor (to the right) and a
linear streaker (center) are shown attached to the tower with weatherproof

Figure 2.10 Streaker and impactor samples attached to a tower as part of a
sulfate study at the General Motors Proving Ground.

covers in place. Both samplers have air intake vertically upward, and their associated diaphragm-type vacuum pumps are located on the ground several meters downwind.

Results obtained with proton scattering analysis on one of the streaker samples are shown in Figure 2.11. Each datum point is separated by two hours and the straight lines are drawn connecting adjacent points as a guide to the eye. The soil-derived elements calcium-potassium and silicon-aluminum appear to be correlated rather well in time while the curves for the elements sulfur and nitrogen bear a general resemblance. Thus, the time variations of these elements promises to aid in interpretation of such experiments.

Information on the composition of the atmospheric aerosol as a function of particle size is also important for health effects assessment and for determination of the origin and transport of the aerosol. In Figure 2.12, impactor results are presented from a study[6] in Los Alamitos, California. Division of the particles into three size ranges reveals striking summer-winter differences in the sulfur and zinc percentages as determined by PIXE.

A similar analysis of five size ranges ($>$ 4, 4-2, 2-1, 1-0.5, 0.5-0.25 μm aerodynamic diameter) taken in Tallahassee, Florida is shown to exhibit a very significant increase in the smallest particle concentration of lead (Figure 2.13). Still another example from a site in Apalachicola National Forest is shown in Figure 2.14. The sulfur concentration decreased during the night with a deep minimum in the particles impacted on stage 3 (aerodynamic diameter 0.5-1.0 μm). With an insufficient number of particle size ranges, the above interesting effects would have gone unnoticed or their magnitudes would have been greatly diminished.

CONCLUSION

Proton-induced X-ray emission and proton elastic scattering analyses are powerful broad-range tools for determination of the elemental content of aerosols. Simple, easily deployed samplers have been developed to facilitate field studies of both time variations and particle size distributions.

ACKNOWLEDGMENTS

This research was supported in part by the U.S. Environmental Protection Agency Grants R-803913 and R-802132.

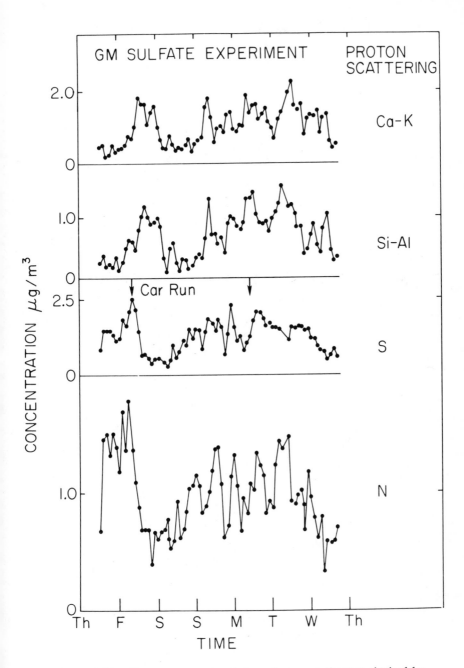

Figure 2.11 Time sequence concentrations for several elements obtained by proton scattering from a streaker sample.

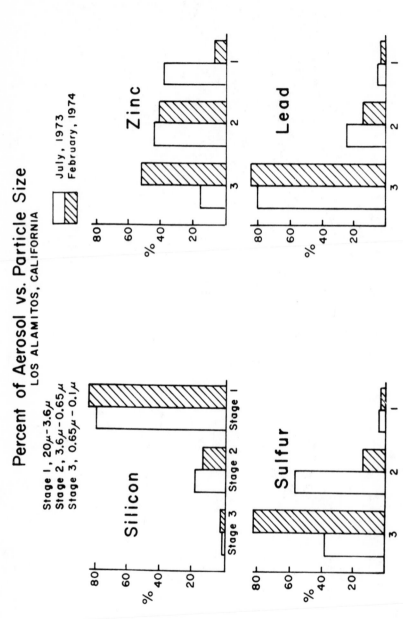

Figure 2.12   Impactor results from a study in Los Alamitos, California, showing marked differences in winter-summer months for sulfur and zinc.

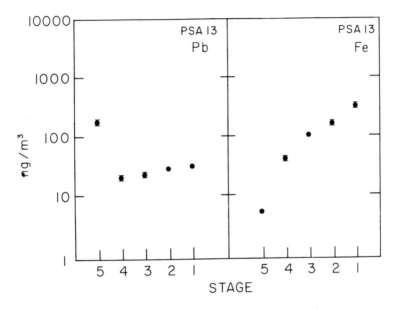

Figure 2.13 Impactor data taken in Tallahassee, Florida, showing smallest particle concentration of lead on stage 5 (aerodynamic diameter 0.5-0.25 μm).

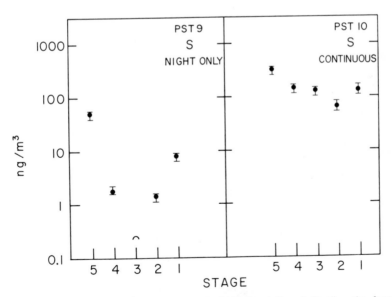

Figure 2.14 Impactor data from Apalachicola National Forest showing the decrease in sulfur concentration at night with a deep minimum at stage number 3 (aerodynamic diameter 2.0-1.0 μm).

REFERENCES

1. Morgan, I. L. *IEEE Trans. Nucl. Sci.* NS-20:36 (1973).
2. Mitchell, R. I. and J. J. Pilcher. *Ind. Eng. Chem.* 51:1039 (1959).
3. Jensen, B. and J. W. Nelson. "Nuclear Methods in Environmental Research," CONF-740701 (Oak Ridge, Tennessee: USERDA Technical Information Center, 1974), p. 366.
4. Kaufmann, H. C., K. R. Akselsson and W. J. Courtney. "REX: A Computer Program for PIXE Spectrum Resolution of Aerosols," in *Advances in X-Ray Analysis*, Vol 19, R. W. Gould, C. S. Barrett, J. B. Newkirk and C. O. Ruud, Eds. (Dubuque, Iowa: Kendall/Hunt Publishing Co., 1976), pp. 355-366.
5. Akselsson, K. R., J. W. Nelson and J. W. Winchester. *Nuclear Cross Sections and Technology*, Vol. 1, NBS Special Publication 425 (Washington, D.C.: U.S. Government Printing Office, 1975), p. 484.
6. Flocchini, R. G. *et al. Environ. Sci. Technol.* 10:76 (1976).

3

SIMULTANEOUS MULTIWAVELENGTH SPECTROMETER
FOR RAPID ELEMENTAL ANALYSIS
OF PARTICULATE POLLUTANTS

Jack Wagman, Roy L. Bennett and Kenneth T. Knapp

Emissions Measurement and Characterization Division
Environmental Sciences Research Laboratory
Research Triangle Park, North Carolina

INTRODUCTION

The chemical characterization of particulate pollutants in the environment has been greatly facilitated by the development in recent years of a number of instrumental multielement analytical procedures. The four techniques used most widely for elemental analysis of airborne particulate samples are optical emission spectroscopy, atomic absorption spectroscopy, neutron activation and X-ray fluorescence. Each of these methods has definite advantages and disadvantages, though none is applicable to all elements.

The use of X-ray fluorescence is noticeably increasing because a number of attributes make it especially attractive for the analysis of airborne particulate matter. These features include (1) the direct analysis of filter deposits with no need for sample preparation, (2) the nondestructiveness of the method, which permits samples to be retained for further analysis or future reference, (3) the fairly uniform detectability across the periodic table, with the ability to analyze all elements from atomic number 9 (F) upward, and (4) the availability of commercial instruments that permit the analysis of samples for a large number of elements in relatively short intervals and at low cost.

Competition exists not only between X-ray fluorescence and other analytical methods but also within the X-ray fluorescence method itself, *e.g.,* between wavelength dispersion and energy dispersion. Birks and his co-workers at the Naval Research Laboratory carried out an EPA-supported laboratory study[1,2] to develop X-ray fluorescence as a method for routine multielement analysis of filter-deposited particulate samples. The study compared various X-ray fluorescence techniques, including the options available for both excitation and detection of the fluorescence. Among the main conclusions were the following:

1. The single-element limit of detection, where interferences are absent or negligible, is about the same for either wavelength or energy dispersion.

2. Atmospheric particulate samples and samples from power plants, incinerators, and other source emissions typically contain many elements at widely different concentrations. Energy-dispersive X-ray fluorescence spectra for such samples show significant interferences between neighboring elements, particularly the elements from sulfur to nickel in the periodic table, and require mathematical unfolding to determine the X-ray intensities. For such real pollution samples, the use of wavelength dispersion spectrometers with their high-resolution capability is a distinct advantage; their use requires considerably less data manipulation, thus avoiding what can be in many instances a major source of error.

3. For routine analysis of large numbers of samples, in which elements of interest can be specified in advance, the use of multichannel wavelength spectrometers appears as the most practical solution, inasmuch as these instruments combine two important features, *i.e.,* high spectral resolving power and simultaneous measurement of a large number of elemental concentrations.

An instrument of this kind, adapted for analysis of filter-deposited samples of particulate matter, has been set up and is in routine use at the EPA Environmental Research Center in North Carolina. Its essential features and performance are described in this chapter.

GENERAL DESCRIPTION OF INSTRUMENT

The high-resolution X-ray fluorescence multispectrometer currently in use at the EPA Environmental Sciences Research Laboratory was set up to provide a means for rapid and quantitative multielement analysis in research projects to characterize airborne particulate samples and to demonstrate its capability for processing large numbers of air pollution samples for routine monitoring purposes. An overall view of the instrument is shown in Figure 3.1. Based upon a newly designed Siemens (Model MRS-3) simultaneous spectrometer* adapted for use with

---

*Simultaneous X-ray multispectrometers are currently manufactured also by Applied Research Laboratories, Philips Electronic Instruments, and Rigaku Denki. Mention of trade names or commercial products does not constitute endorsement or recommendation for use.

Figure 3.1   Simultaneous X-ray multispectrometer system.

filter-deposited samples, the instrument includes automatic sample handling and computer-controlled operation and data processing capabilities.

The EPA simultaneous spectrometer has been equipped with 16 fixed monochromators and 1 scanning monochromator. The fixed channels permit each element from fluorine upward in atomic number to be determined by an optimally designed monochromator, including an appropriate detector for the analyte X-ray energy, and an optimum crystal. The crystals in the fixed channels are logarithmically bent in order to achieve a constant Bragg angle over the entire surface without grinding.[3,4] A list of the elements selected for the fixed channels is shown in Table 3.1 along with the analyte lines, crystals, and detectors employed for each channel.

Table 3.1   Fixed Monochromators in EPA Spectrometer

Electronic Channel	Element	Line	Detector Type	Crystal	Window Thickness, μm	Spectrometer Position
1	Chromium	K_α	Scintillation	LiF(200)	–	1
2	Lead	L_β	Scintillation	LiF(200)	–	1
3	Manganese	K_α	Scintillation	LiF(200)	–	2
4	Arsenic	K_β	Scintillation	LiF(200)	–	3
5	Mercury	L_α	Scintillation	LiF(200)	–	5
6	Bromine	K_α	Scintillation	LiF(200)	–	9
7	Phosphorus	K_α	Flow	PET	2	10
8	Silicon	K_α	Flow	PET	2	2
9	Cadmium	L_α	Flow	PET	6	3
10	Aluminum	K_α	Flow	PET	2	4
11	Sulfur	K_α	Flow	PET	2	5
12	Sodium	K_α	Flow	KAP	0.4	7
13	Fluorine	K_α	Flow	KAP	0.4	4
14	Magnesium	K_α	Flow	ADP	0.4	8
15	Potassium	K_α	Flow	PET	6	8
16	Chlorine	K_α	Flow	PET	6	9

The EPA instrument is also equipped with a fully focusing curved-crystal[5] scanning spectrometer, which greatly increases the flexibility of operation and the total number of elements that may be analyzed. Scanning is accomplished by a stepping motor that may be operated manually or automatically according to predefined scan plans that have been inserted into the minicomputer program.

Fluorescence excitation in the EPA spectrometer is accomplished by either of two interchangeable Siemens type AG61 water-cooled X-ray tubes. Currently, both chromium and rhodium target tubes are used. The high-voltage generator for the X-ray tube has a maximum output of 4 kW. It provides for the adjusting, reading, and stabilization of the tube voltage and current. Analyses are generally carried out with the chromium target tube, which is operated at 50 kV and 52 ma. When the chromium target tube is used, analyses of samples for chromium are done with automatic interposition of an aluminum filter in the primary beam.

The detector amplifier signals from each of the 17 monochromators in the EPA simultaneous spectrometer are processed by a separate compact electronic module. Each of the 17 compact electronic channels consists of a linear amplifier, a discriminator, and a pulse counter. Two types of electronic modules are used—one for scintillation detectors and the other for proportional flow detectors. The input pulses are processed in four three-stage amplifier sections with adjustable gain. The pulse-height discriminator allows only those pulses that lie between the upper and lower voltages of the "channel width" to arrive at the output. These pulses are counted by an eight-decade capacity counter during a given, preset time interval.

The characteristic X-rays of light elements are significantly reduced by absorption in air so that vacuum operation is used to analyze the lower atomic number elements. The X-ray tube, specimen holder, and array of 17 spectrometers are enclosed in a constant temperature spectrometer tank. A vacuum control system provides a fully automatic sequence once the start button is pushed. The sample is evacuated in a forechamber, then inserted into the evacuated spectrometer tank, counted, removed from the spectrometer tank, and returned to atmospheric pressure as the vacuum is released.

A custom-designed automatic sample loader has been fabricated to feed filter samples to the inlet port of the MRS-3 spectrometer and remove the samples after they have been analyzed. The sample loader holds up to 100 filter samples mounted in EPA-designed plastic frames.

A Digital Equipment (DEC) PDP-11/05 minicomputer with a 28K-word memory is used to control the entire spectrometer operation, including insertion and removal of samples; counting of samples by the fixed-channel monochromators for programmed time intervals; programmed operation of the scanning channel for selected elements; storing of calibration and inter-ference correction factors; and handling of output data from all the fixed channels and the scanner. A DEC RT-11 real-time magnetic tape system is used to expand the capacity of the computer. Print-out of the results is made on a Silent 700 Texas Instrument thermal printer.

SPECTROMETERS

The monochromators are located above the specimen and opposite the X-ray tube within the vacuum-tight spectrometer housing. Arranged in a semicircle about the specimen, the monochromators are mounted at ten positions. At seven of these positions, a double monochromator is used, while two single monochromators and the scanner occupy the remaining positions.

As shown schematically in Figure 3.2 each monochromator contains an inlet slit, a logarithmically bent crystal, and an outlet slit in a focusing arrangement. Both the entrance and exit slits are adjustable in direction and width. Fine adjustment of the reflecting angle (θ) by rotation of the

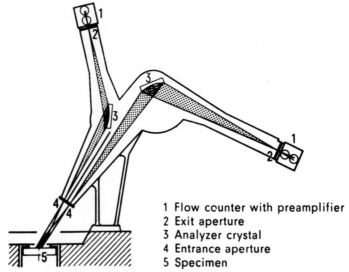

1 Flow counter with preamplifier
2 Exit aperture
3 Analyzer crystal
4 Entrance aperture
5 Specimen

Figure 3.2 Beam paths in a double monochromator.

crystal holder and adjustment of the 2θ angle by the movement of the exit slit is possible after the monochromators are installed, the vacuum applied, and thermal equilibrium attained. This is accomplished by the attachment of a remotely controlled device to two spindles that drive gears that rotate the crystal holder and move the exit slits.

The scanning channel (Figure 3.3) is a curved-crystal spectrometer adjustable over a 2θ range of $30°$ to $120°$, and is compactly designed so that it occupies only one spectrometer position. A fully focusing spectrometer, the scanner is designed so that the inlet slit, crystal, and detector slit are always on the Rowland circle as indicated in Figure 3.4. The sequence of scanner photographs in Figure 3.3 shows that this is accomplished by three guide arms of equal length having bearings so that they can rotate with a common shaft at the Rowland circle center, which they connect to the inlet slit, the crystal, and the detector slit, respectively. A guide band ensures that the distances between the crystal and the two slits are always equal.

Since elements of atomic numbers 9 through 19 (fluorine through potassium) were included for measurement by fixed channels, the scanning spectrometer is equipped with a LiF (200) crystal, thus making it effective for analysis of all higher atomic number elements. Figure 3.5 shows the analytical range of the LiF crystal, and of others that might be selected, and some of the characteristic lines used for analysis. For the effective energy range covered by the LiF crystal, a sealed proportional counter serves as the most efficient detector.

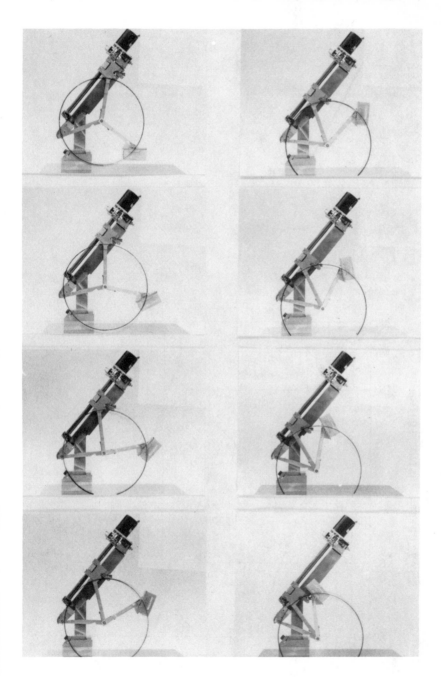

Figure 3.3 Sequence of views of scanner at eight wavelength settings (the Rowland circle was sketched on the photographs to show its shift with wavelength).

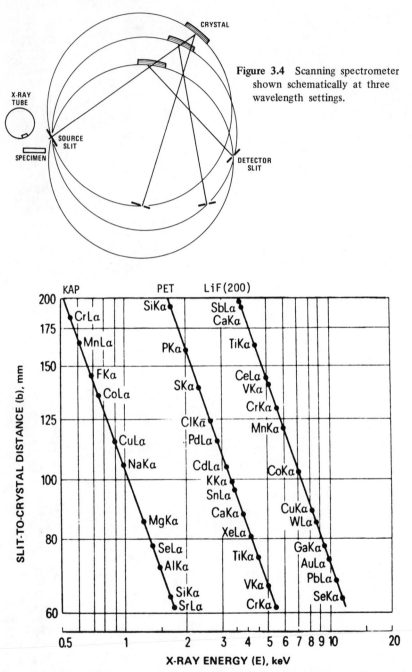

Figure 3.4 Scanning spectrometer shown schematically at three wavelength settings.

Figure 3.5 Scanner slit-to-crystal distance as a function of X-ray energy.

Resolution of a spectrometer is defined as its capability for separating two spectral lines of nearly the same wavelength. The resolution of a crystal spectrometer may be expressed in terms of the full width at half maximum (FWHM) of the diffracted characteristic lines. Table 3.2 lists resolutions measured at four spectral lines with the scanning LiF crystal spectrometer and with a state-of-the-art energy dispersion spectrometer, respectively. For the most difficult series of analyte lines to separate,

Table 3.2   Resolutions Measured with Scanning Wavelength and Energy Dispersive Spectrometers

			Resolution, FWHM		
			LiF (200) Crystal Spectrometer		Energy Dispersive Spectrometer[a]
Element Line	2θ, deg	Energy, keV	$\Delta\theta$, deg	ΔE, eV	ΔE, eV
Pb L_α	33.93	10.55	0.26	154	185
Zn K_α	41.78	8.63	0.22	87	170
Mn K_α	62.97	5.90	0.20	33	155
Ca K_α	113.09	3.69	0.19	6	125

[a]Princeton Gamma-Tech Model PGT-1000.

i.e., the K_α and K_β lines for atomic numbers 20 through 30 (Ca to Zn), the EPA scanning crystal spectrometer has a resolution capability from two- to twenty-fold greater than that of the energy-dispersive solid state detector. Figure 3.6 compares the spectrum obtained for a $CaCl_2$-KCl mixture by the scanning channel with that obtained on the same sample by the energy-dispersive spectrometer. The K_β potassium line, which is easily separated by the scanning channel, is evident in the energy-dispersive spectrum only as a slight distortion in the right side of the calcium K_α peak. The resolution that can be achieved by the scanner for the K lines of Mn, Fe and Co is shown in Figure 3.7.

SAMPLE CONFIGURATION AND HANDLING

Filters and Sample Deposits

Filters used to collect air pollution particulate samples for X-ray analysis should have the following characteristics: (1) low mass to reduce scattering of incident radiation, hence minimizing the background count, (2) low

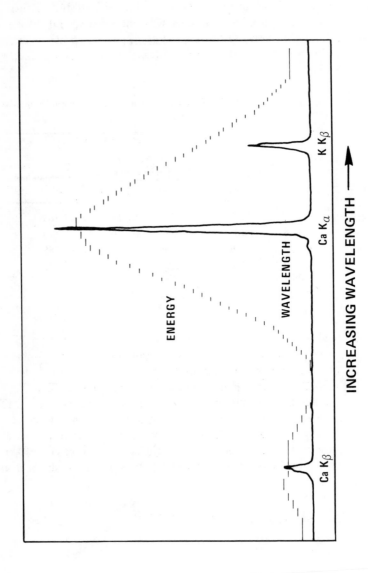

Figure 3.6 Comparison of X-ray spectra obtained with energy and wavelength spectrometers on a calcium-potassium sample.

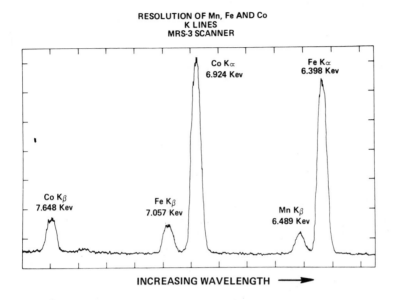

Figure 3.7   Separation of K lines of Mn, Fe and Co by the scanning channel.

content of the elements to be measured, and (3) minimal penetration of particles below the surface.   Also, filters composed of low atomic number elements are desirable since these, in general, have lower attenuation coefficients and therefore lower absorption effects when penetration of the aerosol into the filter occurs.   Filter materials that have been found most suitable for X-ray fluorescence analysis are the thin membrane types, *e.g.*, Millipore, Fluoropore and Nuclepore.

It is advantageous to use thin aerosol deposits for X-ray fluorescence analysis, inasmuch as matrix effects requiring corrections for attenuation and enhancement are thereby greatly reduced.   The degree to which the matrix problem is eliminated depends on the element to be analyzed. In general, thinner samples are required for the lower atomic number elements.   The criterion for a sufficiently thin sample as suggested by Rhodes[6] is:

$$ m < \frac{0.1}{\bar{\mu}} $$

where m is mass deposit density in g/cm^2 and $\bar{\mu}$ is the mean mass attenuation coefficient in cm^2/g.   In a matrix of typical flyash composition,

analyses for sulfur (K_α) and iron (K_α) would require deposit densities
not exceeding about 80 and 1200 $\mu g/cm^2$, respectively. However, fluorine,
which has a $\bar{\mu}$ value in the same matrix of about 9000 cm^2/g, would require a total deposit of approximately 10 $\mu g/cm^2$ or less to meet the
thin-film criterion. When multielemental analyses are to be carried out,
it is obvious that compromises will have to be made in selecting optimal
deposit densities, and matrix correction factors will often be required for
the lighter elements.

Sample Handling

To facilitate the processing of sample deposits in the multichannel
analyzer, special two-piece plastic frame mounts were designed for 47-mm
filters (Figure 3.8). The mounts consist of a lower ring with a 50-mm
outer diameter and 32-mm inner diameter, and an expansible upper ring
of slightly smaller outer diameter. The lower ring has a lip around the
outer edge that holds the filter in place and retains the upper ring. The
filter is placed in the lower ring with the deposit side up; the upper ring
is contracted with snap ring pliers (which fit into two holes); it is then
placed over the filter and allowed to expand into the beveled edge of the
lower ring.

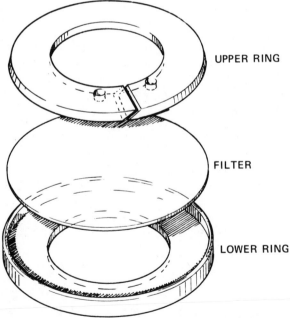

UPPER RING

FILTER

LOWER RING

Figure 3.8   Exploded view of a frame-mounted filter.

When the sample is inserted into the specimen area of the spectrometer, it is pressed against the underside of a gold-plated aperture plate that has a 30-mm diameter opening. This corresponds to the maximum sample deposit area exposed to primary X-ray excitation from the tube.

Frame-mounted filter samples are transported from the external sample loader into the vacuum tank of the spectrometer while seated atop specially designed sample cups. The sample cup serves as a trap for the primary beam that penetrates the sample and is gold-plated to reduce the quantity of backscattered X-rays.

CALIBRATION

Standards are required for calibration of the spectrometer for each element. The sensitivities in terms of counts per second per unit surface deposit concentration, *i.e.,* cps per $\mu g/cm^2$, are determined and stored in the computer memory. As shown earlier, line interferences are minimal with wavelength dispersion; however, in those cases where they do occur, the interference factors are measured using pure standard specimens and are stored in the computer memory.

Several types of thin deposited standards have been employed to calibrate the spectrometer. These include:

1. Thin films formed from vacuum-evaporated elements and compounds. These were prepared on Mylar substrate (3.8-μm thickness) by Micromatter Company at concentrations of about 50 $\mu g/cm^2$ with an estimated accuracy of 5%.

2. Aerosols generated, *e.g.,* in a Collison atomizer, and collected on membrane filters.

3. Filter deposits of materials from solutions after solvent evaporation. This method is subject to nonuniform deposits caused by migration of ions during evaporation unless precautions are taken.

4. Filter deposits of fine powdered materials from liquid suspension.

The data presented in the remainder of this chapter are based upon the use of vacuum-evaporated films as standards.

INSTRUMENT OPERATION AND DATA PROCESSING

Instrument Control

The operation of the fixed channels is controlled by the selection of any one of a number of possible programs hardwired on program cards. These permit the simultaneous analysis of up to 16 selected elements for any desired counting time interval. As counting in the fixed channels begins, operation of the scanner can also proceed automatically as a result of the selection of one of several scan plans that have previously been defined and stored in the computer memory. A given scan plan contains the elements to be determined on the sequential channel and the counting time interval for each. The program also allows the operator the option of selecting a maximum count (*e.g.,* 10,000) for each element to reduce the total count time when one or more elements are present at high concentration. In this option, the X-ray intensity from each element is measured automatically for either the designated time interval or the maximum count, whichever occurs first.

Calibration, Measurement and Data Processing

Before elemental concentrations in particulate samples can be determined, a set of calibration standards must be inserted and processed in the instrument. The computer-controlled calibration program includes measurement of standards at the analyte lines and at other lines at which the material in the standard may interfere. Sensitivity factors for all of the elements and interelement line interference ratios are calculated and stored in the computer memory.

Analysis of a series of particulate samples on a given filter substrate is preceded by measurement of a corresponding filter blank for automatic determination and storage of background count levels and minimum detection limits for each of the elements selected for measurement. Following measurement of each particulate sample, the computer program subtracts background counts, makes line interference corrections, and applies sensitivity factors to convert cps to $\mu g/cm^2$. The list of elemental concentrations for each sample is stored on magnetic tape for later reference, or is printed out at the data terminal, or both; those element concentrations that are below detection limits are noted by asterisks. The program also provides for the application of other correction factors as needed, *e.g.,* for variations in particulate deposit area and particle size.

INSTRUMENT PERFORMANCE

Sensitivities and Detection Limits

The sensitivity of the X-ray spectrometer for a given element is defined as the rate of change in the analyte-line intensity with a change in the analyte concentration. The minimum detection limit for X-ray analysis is often defined as the concentration of analyte corresponding to a net count equal to three times the standard deviation, *i.e.,* three times the square root of the background count.   The measured sensitivities and detection limits for 100-second count times on 30 elements determined in the EPA simultaneous spectrometer, with thin-film-on-Mylar calibration standards, are listed in Table 3.3   It is noteworthy that the minimum detection limits for only 100 seconds of count time are in the range of 2 to 10 ng/cm^2 for more than half of the elements measured including most of the low atomic number elements.   The detection limit exceeds 30 ng/cm^2 for only four of these elements.

Table 3.3   EPA Multispectrometer XRF Analyzer Element Sensitivities
and Detection Limits

Element	Sensitivity, Counts/100 sec $\mu g/cm^2$	Detection Limit (100 sec, 3σ), ng/cm^2	Element	Sensitivity, Counts/100 sec $\mu g/cm^2$	Detection Limit (100 sec, 3σ), ng/cm^2
F	220	149	Co	16540	3
Na	534	29	Ni	14504	10
Mg	10280	2	Cu	18880	43
Al	8074	3	Zn	21066	7
Si	11614	3	As	17125	10
P	13392	15	Se	22922	12
S	28013	9	Br	50340	28
Cl	25394	9	Cd	17303	2
K	121286	2	Sn	14800	2
Ca	87817	2	Sb	31100	4
Ti	85635	2	Ba	25000	7
V	18010	7	Pt	6812	20
Cr	7484	19	Au	8498	91
Mn	17522	14	Hg	5776	90
Fe	13300	18	Pb	16583	30

Precision and Accuracy Estimates

The precision of X-ray fluorescence analysis depends on a combination of factors, including statistical counting error, instrument drift, specimen variation, and miscellaneous operational errors.   In order to evaluate the magnitude of these factors, the replicate determinations listed in Table 3.4 were conducted.   A large accumulated count was used to minimize the relative standard deviation resulting from statistical counting error. A single sample counted in place ten times showed a relative standard deviation of only 0.05%, while the count for a sample introduced into the instrument ten times had a relative standard deviation of 0.10%.

Table 3.4   Precision of Replicate Analyses of Potassium[a]

	Standard Deviation, σ	Relative Standard Deviation $(\frac{\sigma}{\overline{N}} $ x 100), %
Statistical counting error, $\sqrt{\overline{N}}$	531	0.02
Observed variation of ten replicate counts on single sample in place	1539	0.05
Observed variation of ten replicate counts on a single sample inserted and removed ten times	2729	0.10

[a]Average count, $\overline{N}$: 2,820,092.

The latter demonstrates the excellent repeatability of the total counting procedure, which includes specimen handling and positioning.

Another factor that must be considered in the overall precision of a measurement is the uniformity of the filter substrate used to collect particulate samples, since a blank filter count is used to correct for background. Table 3.5 lists the average background counts and standard deviations for ten blank specimens of each of three types of membrane filters frequently used. Nuclepore filters exhibited the best overall uniformity.

An estimate of the accuracy of the X-ray analytical procedure, as carried out with the simultaneous spectrometer, was determined in a comparison with gravimetric values for the mass of potassium sulfate aerosol generated in a Collision atomizer and deposited on a series of Nuclepore filters. The results plotted in Figure 3.9 show very good agreement between the X-ray and gravimetric values. Another comparison, this time with atomic absorption analyses, is shown in Table 3.6. The samples consisted of particle emissions collected on high-purity quartz filters from a controlled combustion source fueled with oil spiked with known amounts of organometallic compounds. The calculated values shown in the table for the five metals analyzed are based upon the amount of spiked fuel consumed and the known fraction of total emissions collected on the filter. The X-ray fluorescence concentrations are in remarkably good agreement with the calculated values with the exception of the cadmium value, which is about 10% low, probably because of some attenuation of the relatively low-energy cadmium fluorescence resulting from only partial penetration of collected aerosol into the interstices of the quartz fiber filter.

Table 3.5   Magnitude and Uniformity of Background Count for Three Types of Membrane Filters

	Nuclepore 0.8μ (Mass = 1.1 mg/cm^2)		Fluoropore Type FA (Mass = 2.7 mg/cm^2)		Millipore Type AA (Mass = 5.0 mg/cm^2)	
	$\overline{N}$[a]	σ[b]	$\overline{N}$	σ	$\overline{N}$	σ
F	1.32	0.09	549.78	28.20	1.96	0.13
Na	0.22	0.06	0.15	0.05	0.28	0.05
Mg	0.64	0.08	0.75	0.10	1.43	0.16
Si	66.22	4.44	1.25	0.25	2.29	0.27
P	17.77	0.62	17.63	2.16	16.34	0.35
S	48.54	0.82	40.21	2.23	54.52	0.63
Cl	86.49	3.18	54.44	4.81	200.24	5.59
K	64.48	2.67	63.49	2.46	212.40	2.40
Ca	4.43	1.17	8.85	3.15	333.84	7.13
Ti	2.91	0.56	7.67	1.20	12.74	1.62
V	7.95	0.80	16.10	1.93	25.71	1.65
Fe	44.28	4.21	50.41	4.86	62.51	4.79
Ni	2.10	0.36	3.51	0.67	5.52	1.12
Cu	553.46	9.03	567.36	14.13	577.69	21.61
Zn	12.14	0.81	26.31	3.49	41.34	6.82
Cd	1.87	0.08	2.12	0.11	3.18	0.17
Ba	2.35	0.35	5.00	0.57	8.27	0.94

[a]$\overline{N}$ is the mean value in cps for ten blank specimens.
[b]σ is the standard deviation in background for ten blank specimens.

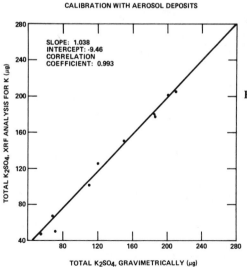

CALIBRATION WITH AEROSOL DEPOSITS

SLOPE: 1.038
INTERCEPT: -9.46
CORRELATION
COEFFICIENT: 0.993

TOTAL K$_2$SO$_4$, XRF ANALYSIS FOR K (μg)

TOTAL K$_2$SO$_4$, GRAVIMETRICALLY (μg)

Figure 3.9 Comparison of X-ray fluorescence and gravimetric determinations of K$_2$SO$_4$ deposited on Nuclepore filters.

Table 3.6   Comparison of XRF and AAS Analyses of Particulate Samples
from a Simulated Combustion Source

	Total Collected, μg		
Element	Calculated Value[a]	Atomic Absorption	X-Ray Fluorescence
Pb	202	228	212
Mn	182	195	183
Co	182	193	183
Cd	183	187	166
V	216	159	230

[a]Based on analysis of fuel oil spiked with organometallic compounds.

Analyses of Source and Ambient Air Samples

The EPA simultaneous wavelength X-ray spectrometer is being used to determine the elemental composition of large numbers of particulate samples collected from mobile sources that include catalyst-equipped cars, diesel engines, and aircraft;   from stationary sources that include power plants, incinerators and chemical process plants; and from the ambient air.   Table 3.7 lists elemental concentrations found in emissions from 1975 cars equipped with oxidation catalysts.   The results show sulfur

Table 3.7   Typical XRF Analyses of Particulate Emissions from 1975
Catalyst-Equipped Cars

	Concentration, $\mu g/cm^2$		
Element	Sample No. 4	Sample No. 8	Sample No. 30
Al	–	0.12	0.05
Si	0.40	0.02	0.11
P	0.05	–	–
S	6.80	23.00	40.50
Ca	0.08	0.08	0.04
Ti	0.02	0.01	0.01
Fe	0.32	1.10	1.40
Ni	0.02	–	0.03
Zn	0.07	0.07	0.10
Ba	0.01	–	0.02
Mass loading	46.40	143.60	270.90

to be the predominant element and are in good agreement with independent measurements indicating that nearly all of the particulate matter consists of droplets of 45% (w/w) sulfuric acid in water.

Table 3.8 shows typical analyses obtained from samples of particulate emissions from a coal-fired power plant, and Table 3.9 lists element concentrations in urban air particulate samples collected at different locations and times. These data illustrate the broad range of elements and elemental concentrations present in air pollution samples.

Table 3.8   XRF Analysis of Flyash from Coal-Fired Power Plant

Element	Concentration, $\mu g/cm^2$		Element	Concentration, $\mu g/cm^2$	
	Sample No. X5	Sample No. X10		Sample No. X5	Sample No. X10
F	0.29	0.15	V	0.55	0.25
Na	0.42	0.35	Cr	23.	15
Mg	2.0	1.2	Fe	20.2	9.2
Al	103.	48.	Ni	0.03	–
Si	55.	26.	Zn	4.6	0.13
P	0.54	1.0	Br	0.89	–
S	7.8	7.8	Cd	0.020	0.007
K	15.6	5.7	Ba	0.62	0.33
Ca	–	4.9	Pb	1.7	1.1
Ti	3.6	1.5			

CONCLUSIONS

The performance of an X-ray fluorescence multispectrometer system has shown it to be a highly effective means for rapid and routine multielemental analysis of large numbers of air pollution particulate samples. Rapid determinations were accomplished by simultaneous analyses with 1 scanning and 16 fixed spectrometers and by the relatively short counting times possible with wavelength dispersive spectrometers.

The higher resolution obtained with crystal spectrometers compared to energy-dispersive instruments reduces the interferences that occur between elements having characteristic lines of nearly the same energy. This is especially important with environmental samples that typically contain several dozen elements, including many requiring interference corrections that are large and uncertain in low-resolution instruments. The use of crystal spectrometers with their high-resolution capability thus minimizes,

Table 3.9   XRF Analysis of Ambient Air Particulate Samples

Element	Concentration, $\mu g/cm^2$		
	Sample 7-19	Sample 7-24	Sample 10-04
F	–	–	0.09
Na	–	–	0.09
Mg	0.02	0.06	0.14
Al	0.28	1.20	5.30
Si	0.44	1.60	7.60
P	0.11	–	0.18
S	3.74	3.20	1.77
Cl	0.26	0.15	0.47
K	0.20	0.47	2.06
Ca	1.10	0.90	2.67
Ti	0.04	0.08	0.16
V	0.005	–	0.03
Fe	0.76	0.70	2.73
Zn	1.10	0.05	0.16
Br	0.40	–	0.75
Cd	0.04	0.007	0.04
Ba	–	0.04	0.10
Pb	1.80	0.23	3.0

and in most cases eliminates, the need for mathematical unfolding procedures that can be a significant source of error in X-ray fluorescence analysis.

The measured sensitivities and detection limits are more than adequate for most elements of environmental concern. For 30 elements of greatest interest the minimum detection limits for a counting time of 100 seconds were in the range 2 to 10 ng/cm^2 and exceeded 30 ng/cm^2 for only four of the elements.

The precision of the analysis depends on counting statistics, instrument drift, specimen variation, and operational errors. Excellent precision in the total counting procedure, including specimen handling and positioning, was found in repeatability tests. Evaluation of the uniformity of various membrane filter substrates revealed that Nuclepore was the most uniform.

Comparison of the results of X-ray fluorescence analyses with the results obtained on the same samples by gravimetric or atomic absorption analysis has verified the accuracy of the method. The analytical capability of the instrument has been demonstrated with a large variety of sample types from mobile sources that included catalyst-equipped cars, diesel engines, and aircraft; from stationary sources that included incinerators, power plants, and chemical process plants; and from ambient air sampled at various locations.

The cost of analyses carried out with the simultaneous X-ray spectrometer has been calculated based upon a 5-year amortization of the capital cost, typical operating expenses, and sample burdens well within the capacity of the instrument. Estimates range from 2 to 4 dollars per sample and as low as 10 cents per element.

ACKNOWLEDGMENTS

The authors are pleased to acknowledge the valuable contributions of L. S. Birks, J. V. Gilfrich and J. W. Criss of the Naval Research Laboratory, Washington, D.C., in the development of specifications for this instrument and in optimization of its performance. We also thank R. B. Kellogg, Northrop Services, Inc., Research Triangle Park, North Carolina, for his excellent assistance in instrument operation and data acquisition.

REFERENCES

1. Birks, L. S., J. V. Gilfrich and P. G. Burkhalter. "Development of X-Ray Fluorescence Spectroscopy for Elemental Analysis of Particulate Matter in Atmosphere and in Source Emissions," U.S. Environmental Protection Agency, Office of Research and Development, Publication No. EPA-R2-72-063 Research Triangle Park, North Carolina, (November 1972).

2. Birks, L. S. and J. V. Gilfrich. "Development of X-Ray Fluorescence Spectroscopy for Elemental Analysis of Particulate Matter, Phase II: Evaluation of Commercial Multiple Crystal Spectrometer Instruments," U.S. Environmental Protection Agency, Office of Research and Development, Publication No. EPA-650/2-73-006, Research Triangle Park, North Carolina (June 1973).

3. Barraud, J. "Monochromateur-focalisateur donnant un faisceau d'ouverture notable," *C. R. Acad. Science*, Paris 214:795 (1942).

4. DeWolff, P. M. "An Adjustable Curved Crystal Monochromator for X-Ray Diffraction Analysis," *Appl. Sci. Res.* B1:119 (1950).

5. Johansson, T. "Über ein Neuartiges, genau Fokussierendes Röntgenspektrometer," *Zeitschrift für Physik.* 82:507 (1933).

6. Rhodes, J. R., A. Pradzynski, R. D. Sieberg and T. Furuta. "Application of a Si(Li) Spectrometer to X-Ray Emission Analysis of Thin Specimens," in *Low-Energy X- and Gamma-Ray Sources and Applications*, C. A. Ziegler, Ed. (London and New York: Gordon and Breach, 1971), p. 317.

4

WAVELENGTH DISPERSION

L. S. Birks

Naval Research Laboratory
Washington, D.C.

INTRODUCTION

The deliberate purpose of this chapter is to emphasize the advantages of wavelength dispersion even though the author is well aware of the merits of energy dispersion for some applications in pollution analysis. The chapter does not include a discussion of the theory, instrumentation, or detailed techniques of X-ray analysis for newcomers; rather, it is addressed to those already familiar with the background and capabilities of the X-ray method but who are interested in where and how wavelength dispersion may be superior to energy dispersion.

Three primary topics will be discussed: (1) the superior resolution of wavelength dispersion allows better selectivity for low-concentration elements and even allows valence states in compounds to be identified, (2) the count rate, which determines sensitivity and limit of detection, is limited to 20,000 cps for the whole spectrum in energy dispersion but can be 50,000 cps for each element in wavelength dispersion, and (3) the gas-proportional detector has advantages over Si(Li) detectors in that it can discriminate in favor of low Z elements.

RESOLUTION

The best resolution claimed for energy dispersion is about 150 eV, which means there is serious overlap of lines from neighboring elements, particularly the $K\beta$ line of element Z from the $K\alpha$ line of element Z + 1,

57

as well as similar problems in the L series lines. Energy dispersion advocates claim that the line interference can be accounted for by mathematical unfolding of the overlapping lines, but this leads to large uncertainties when the line of interest is weak and the interfering line is strong. In wavelength dispersion the best resolution is about 3 eV or better, which eliminates the vast majority of line interferences.

Selectivity

As an example of the effect of resolution on the selectivity for weak lines, consider the case of Cd Lα in aerosols where potassium is present at 50-100 times greater concentration. Figure 4.1 shows an idealized plot in the region of potassium. The dotted curve is for 0.3 μg/cm^2 of potassium at 150 eV resolution by energy dispersion (Gaussian shape

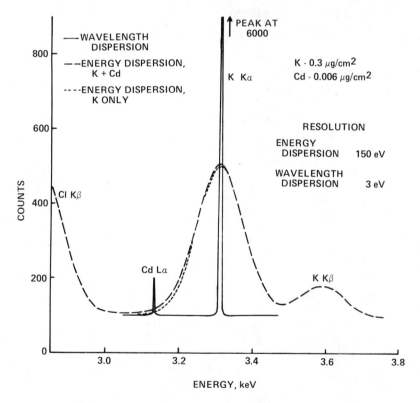

Figure 4.1   Comparative resolution of cadmium and potassium by energy dispersion and wavelength dispersion. (The Cd Lβ occurs within 3 eV of K Kα and is not resolved even with wavelength dispersion.)

assumed). The dashed curve shows the addition of 6 ng/cm^2 of cadmium. Unfolding the weak cadmium peak from the dashed curve would involve considerable error and in practice would not be feasible; therefore energy dispersion would have to measure Cd Kα radiation at which energy the fluorescence/scattered ratio is generally not as favorable. By contrast, the same concentration of cadmium is easily distinguished in the presence of potassium using wavelength dispersion as illustrated by the solid line in Figure 4.1.

Valence

A second advantage of good resolution is the ability of wavelength dispersion to identify different valence states of important elements, such as sulfur. Figure 4.2 shows recent measurements made at the Naval Research Laboratory on 50 μg quantities of sulfate, sulfite and sulfide compounds. The satellites of S Kβ indicate the valence state of sulfur unambiguously. Note that the resolution with the NaCl crystal is about 3 eV as indicated by the energy scale on Figure 4.2.

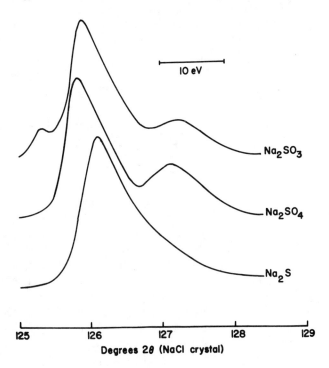

Figure 4.2 Valence band satellites of S Kβ for different classes of sulfur compounds.

COUNT RATE AND LIMIT OF DETECTABILITY

As all X-ray analysts are aware, the count rate for an element line determines the sensitivity for that element, and the count rate for the background determines (through statistics) the limit of detection that may be achieved in a specified counting interval.

In energy dispersion the Si(Li) detector is count-rate limited to 10,000-20,000 counts per second, but this must include all of the sample spectrum plus scattered primary radiation. As a matter of fact, the scattered primary radiation and background can comprise as much as 99% of the total. This leaves very little allowed counts for the elements of interest, especially the minor constituents.

Standard X-ray tubes generally excite too much intensity for the allowed detector count rate and consequently must be operated at greatly reduced power or with secondary fluorescers or filters. Radioisotope sources, on the other hand, are weak emitters as comparison with X-ray tubes but are frequently adequate for energy-dispersion applications. As for proton excitation, there is no scattered primary peak to swamp the detector, but there is a strong background intensity (bremsstrahlung, Compton scattering, etc.) at low energy that can interfere with measurement of the low Z elements.

In wavelength dispersion the gas proportional detector has a general count-rate limitation of about 50,000 per second, but this is seldom if ever achieved for individual elements in pollution samples even at maximum X-ray tube power. Therefore, the only count rate limitation in wavelength dispersion is due to X-ray tube power, and the analyst can use all the intensity available.

The limit of detection is generally defined as a line intensity that exceeds the background by three times the standard deviation of the background under the line (and therefore is directly related to resolution as well as count rate). Table 4.1 shows the limits of detection for sulfur, cadmium and lead for energy dispersion, as presented in recent papers by Giauque et al.[1,2] and, for comparison, current measured values from the EPA wavelength dispersion instrument.[3] The lead value for wavelength dispersion in Table 4.1 reflects interference from the gold sample cup, which could be exchanged for a different sample cup material thereby improving the detection limit considerably.

Without quibbling about whether the single-element limit of detection is comparable by the two methods, one can at least state that X-ray fluorescence by either method is generally sensitive enough for the quantities of elements of interest in air pollution samples.

Table 4.1   Detection Limits (3σ) in ng/cm^3

Element	Energy Dispersion[a]	Wavelength Dispersion[b]
Sulfur	12	9
Cadmium	20	2
Lead[c]	50	30

[a]Estimated from references 1 and 2 for 5 mg/cm^2 Millipore filter and 600 seconds counting time.
[b]From reference 3 for 3.8 μm Mylar (0.5 mg/cm^2) substrate and 100 seconds counting time.
[c]Both values can be reduced by taking suitable precautions.

GAS PROPORTIONAL DETECTORS

The spectacular advent of the Si(Li) detector in the 1960s stopped all attempts to improve gas detectors. Thus it has been generally forgotten that gas proportional counters were used successfully in the 1950s for energy dispersive analysis. It has also been forgotten that gas detectors have one important advantage over solid state detectors: the gas detector can discriminate in favor of low Z elements. This is accomplished by choosing the gas and the pressure that are efficient for absorbing soft X-rays but inefficient for hard X-rays. As a numerical example of this, consider the use of a nickel fluorescer to excite S $K\alpha$ in energy dispersion. The scattered nickel primary radiation comprises over 99% of the total radiation detected by a Si(Li) detector. If a neon gas proportional detector were used instead, it would still be 87% efficient for the S $K\alpha$ but only 6% efficient for Ni $K\alpha$, thereby reducing the primary scattering by a factor of nearly 20.

The problem with gas proportional detectors that has eliminated them from competition with Si(Li) is simply their relatively poor resolution of about 900 eV at Mn $K\alpha$ and 600 eV at S $K\alpha$. But, as has been stated,[4] it is not statistics of the number of initial ion pairs that limits resolution in a gas proportional detector even though that is the term commonly used in the resolution formula. Rather it is the variation in the internal amplification process; there may be some possibility of improving this situation. There has already been a publication that describes a different detector geometry and that claims a factor of two improvement in resolution.[5] Such promising developments make one hope for gas detectors that are more nearly competitive with Si(Li) in resolution and that have better selectivity for low Z elements.

CONCLUSIONS

This chapter has described the significance of the better resolution and count-rate capability enjoyed by wavelength dispersion. From the results shown several statements can be made about the relative uses and merits of wavelength and energy dispersion.

1. Energy dispersion is the best survey instrument for determining major constituents in completely unknown samples.
2. Wavelength dispersion is more capable of measuring low concentrations of elements in the presence of high concentrations of other elements.
3. Once the elements of interest in pollution are specified by EPA, the multiple crystal spectrometer wavelength dispersion instrument can be the most rapid (hence most economical) method for quantitative analysis of large numbers of samples.
4. Only wavelength dispersion offers the possibility of measuring valence and bonding of the elements.

REFERENCES

1. Giauque, R. D., R. B. Garrett, L. Y. Goda, J. M. Jaklevia and D. F. Malone. *Adv. X-Ray Anal.* 19:305 (1976).
2. Giauque, R. D., L. Y. Goda, and R. B. Garrett. University of California at Lawrence and Berkeley Laboratory Report LBL-4414 (October 1975).
3. Wagman, J., R. L. Bennett and K. T. Knapp. Environmental Protection Agency Report EPA-600/2-76-033 (March 1976).
4. Fink, R. W. *Atomic Inner Shell Processes*, Vol. 2, B. Crasemann, Ed. (New York: Academic Press, 1975).
5. Palmer, H. E. and L. A. Braby. *Nuc. Instr. Methods* 116:587 (1974).

5

COMPARISON OF MINIMUM DETECTABLE LIMITS
AMONG X-RAY SPECTROMETERS

Joseph M. Jaklevic

Lawrence Berkeley Laboratory
University of California
Berkeley, California

Richard L. Walter

Department of Physics, Duke University
Durham, North Carolina   and
Triangle Universities Nuclear Laboratory
Duke Station, North Carolina

INTRODUCTION

Methods for applying the various X-ray techniques to air particulate samples have been under development over the past several years at a number of laboratories. In some cases the technology has progressed beyond the experimental prototype stage to the construction of complete analysis facilities capable of processing large numbers of samples on a routine basis. It therefore seems appropriate that the capabilities of the various methods be compared on the basis of these existing facilities.

The comparison is based on measurement of minimum detectable limits for single elements for each of three X-ray fluorescence methods. These limits are derived from measured sensitivities and background counting rates. It is assumed that the background fluctuations are determined solely by random Poisson counting statistics. The peak-to-background ratio achieved at the detectable limit is also derived from these quantities. This parameter is of interest in assessing the susceptibility of the measured concentration to small changes or drift in the background levels such as might be introduced in certain types of spectral analysis.

63

The emphasis on minimum detectable limits as a parameter for comparison does not imply that it is the only governing factor in determining the utility of a given analytical method. The ultimate goal of any analytical measurement is to either determine the concentration of a given element to a desired degree of accuracy or to make a decision whether or not a particular element is present at some specified level. The single element detectabilities quoted in this discussion represent optimistic predictions regarding the statistical accuracy of any such measurement. Possible systematic errors or erroneous background determinations could increase the detectable limits over those quoted. Analysis of realistic samples can further complicate the picture because of interelement interferences, X-ray absorption and enhancement effects, and possible changes in the background characteristics. Each of these affects the various methods in different ways so that a simple comparison can no longer be made.

Three multielement X-ray fluorescence methods were considered: (1) wavelength dispersive analysis using 16 fixed crystals and 1 scanning channel, (2) energy dispersive analysis using a series of discrete energy photon sources and a Si(Li) spectrometer, and (3) energy dispersive analysis using 3-MeV proton excitation and a Si(Li) X-ray spectrometer.

DEFINITION OF PARAMETERS

The expression used for the minimum detectable limit is based on the derivation used by Currie.[1]   Using his Equation 2 we define the detection limits for a given method

$$C_D = 3.29 \ \sigma_0 \qquad\qquad (5.1)$$

where σ_0 is the standard deviation of the observed result. This represents a 95% confidence level of detection at the decision limit C_c, which is defined as[1]

$$C_c = 1.64 \ \sigma_0 \ \cong C_D/2 \qquad\qquad (5.2)$$

If we assume that the standard deviation at the minimum detectable limit is determined solely by the Poisson distributed counting statistics in the background under the X-ray peak, then Equation 5.1 becomes

$$C(mdl) = 3.29 \ (R_b/t)^{\frac{1}{2}}/S \qquad\qquad (5.3)$$

where R_b is the background counting rate in counts/sec under the X-ray peak, t is the time interval of the measurement, and S is the sensitivity of the instrument for that specific element expressed as counts/sec per $\mu g/cm^2$ of concentration.

The peak-to-background ratio at the minimum detectable limit is given by

$$PBR(mdl) = S \cdot C(mdl)/R_b = 3.29/(R_b t)^{1/2} \qquad (5.4)$$

This parameter can be used to appraise the influence of random errors in the measurement of the background. In many forms of X-ray spectral analysis, the background level under a peak is inferred from background measurements over widely separated regions of the spectrum. The possibility exists that this estimate could be slightly in error as the sample form or composition is varied.

In order to include this effect in discussions of minimum detectable limit it is useful to consider an expanded limit as follows:

$$C'(mdl) = C(mdl) \left[1 + \frac{\epsilon}{PBR(mdl)} \right] \qquad (5.5)$$

where ϵ is a fractional uncertainty in the background brought about by systematic errors above the random errors included in $C(mdl)$. If we assume, for example, that the estimated background is 10% less than the actual background, i.e., $\epsilon = 0.1$, subsequent analysis will leave residual concentration for that element equivalent to 10% of the actual background. The value of this false measurement would be equal to the minimum detectable limit if $PBR(mdl) = 0.10$. Since typical $PBR(mdl)$ can range from 0.05 to 2, this illustration emphasizes the importance of valid background subtraction methods for low-level detection.

RESULTS

Data are presented for three specific systems. The first system is a wavelength dispersive XRF unit that contains 16 fixed crystal spectrometers and 1 crystal with scanning capabilities.[2] Excitation is provided by the direct output from Cr or Rh anodes. Wavelength systems are commercially available from a number of manufacturers, and this particular one was purchased from Siemens. At least one company provides a system with about twice as many fixed crystals as the one referred to here.

The second XRF system is the pulsed Lawrence Berkeley Laboratory (LBL) unit for energy dispersive analysis described by Jaklevic.[3] It employs secondary fluorescers of titanium, molybdenum or samarium for generating the exciting radiation. Because the fluorescent radiation from the samples is detected with a Si(Li) detector, the technique possesses all the advantages and the disadvantages of recording and analyzing a continuous spectrum of the X-ray yield. The predecessor to this system was also developed at the Lawrence Berkeley Laboratory and has been described previously.[4] This latter system is not pulsed and operates at lower counting rates (*i.e.,* approximately 5000 counts per second, with minimum detectable limits ranging from 2 to 10 times higher than the newer pulsed system). Commercial energy-dispersive units are also available from several firms, and these units can perform at minimum detectable limits comparable to the older LBL system. At least one significant difference concerning detection limits exists between both LBL systems in comparison to the wavelength system above and the PIXE system described below. The two energy-dispersive XRF systems operate with the samples in a helium atmosphere and thereby avoid the complications (and losses) of dealing with vacuum chambers. However, some of the background counts accumulated for very thin samples can be attributed to the X-ray scattering from the helium.

For the comparison, detectable limits from two particle-induced X-ray emission analysis (PIXE) systems will be considered. These are the 5-MeV and 3-MeV proton-beam systems operated at the Florida State University[5] (FSU) and Duke University,[6] respectively. Since the data were intended to represent optimum single element detectabilities, absorbers were used to reduce the low-energy continuum background in the case of measurement for the elements with higher energy K_α lines. This is analogous to the use of multiple fluorescent X-ray energies in the case of photon-excited fluorescence. In routine analysis a compromise between using absorbers is to employ a leaky absorber, that is, one which has a small hole that permits a few percent of the soft X-rays to reach the Si(Li) detector. The primary advantage is that only one irradiation is necessary per sample, but one pays the price of having a fairly complicated continuous background to take into account.

The results for the three types of systems are presented in Tables 5.1, 5.2 and 5.3 for a range of elements. The sensitivity (in units of counts/ sec per $\mu g/cm^2$) is given in the left hand column and is the same regardless of the substrate on which the material is deposited. The background counts correspond to the counts occurring within the wavelength or energy window selected for the devices for counting or intergrating the "peaks." For the energy dispersion systems, the area of integration is restricted to the central region of the peak, which contains about 70% of the total area, for optimal signal to noise ratio. Values of R_b are included in the tables using the following notation:

R_{bo} = Background count rate (counts/sec) under peak of element Z
when no sample or substrate is present.

R_{bs} = Background count rate (counts/sec) under peak of element Z
when a clean sample substrate is present.

Table 5.1   EPA Wavelength Dispersive System

Element	S	No Membrane			Mylar			Nuclepore			Millipore		
		R_{bo}	$C(mdl)^a$	PBR	R_{bs}	$C(mdl)^a$	PBR	R_{bs}	$C(mdl)^a$	PBR	R_{bs}	$C(mdl)^a$	PBR
Al	130	0.3	1.3	0.63	0.5	1.7	0.49	0.4	1.7	0.50	0.8	2.2	0.38
Si	100	1.1	3.4	0.31	2.4	5.1	0.21	103	33	0.03	2.7	5.4	0.20
S	320	21	4.9	0.07	24	5.1	0.07	36	6.3	0.05	41	6.7	0.05
K	1200	43	1.8	0.05	45	1.8	0.05	45	1.8	0.05	210	3.9	0.02
V	160	1.5	2.4	0.27	4.0	4.0	0.17	9	6.0	0.11	23	9.5	0.07
Fe	140	25	12	0.06	29	13	0.06	34	14	0.06	51	17	0.05
Zn	180	1.1	2.0	0.31	2.6	2.9	0.21	4.5	3.9	0.16	14	6.9	0.09
As[b]	130	5.8	6	0.14	10	7.9	0.11	15	10	0.08	43	17	0.05
As[c]	40	7800	318	0.004	7900	320	0.004	8100	325	0.004	8400	330	0.004
Se	230		12										
Sr													
Cd	180	1.2	2.0	0.31	1.5	2.2	0.27	2.1	2.7	0.23	3.1	3.3	0.19
Ba	220	0.3	0.8	0.60	1.5	1.8	0.27	1.9	2.0	0.24	4.9	3.2	0.15
Sn	140	0.4	1.4	0.56	1.2	2.6	0.30	4.0	4.8	0.16	3.9	4.7	0.17
Pb	160	250	32	0.02	300	35	0.02	390	40	0.02	700	53	0.01

[a] $C(mdl)$ is in units of ng/cm^2.   Data represent 100-second analysis.

[b] Scanning crystal results.   Set for $K\alpha$ line.

[c] Fixed crystal results.   Set for $K\beta$ line.

Table 5.2   LBL Energy-Dispersive System

| Element | S | No Membrane | | | Millipore | | |
		R_{bo}	C(mdl)[a]	PBR	R_{bs}	C(mdl)[a]	PBR
Al[b]	4.0	3.9	160	0.16	25.7	417	0.065
Si	11.2	4.4	62	0.16	28.7	157	0.061
S	63.5	8.3	15	0.12	53.6	38	0.045
K	227	12.3	5.1	0.09	79.7	13	0.037
V[c]	37.3	1.7	11.0	0.24	6.2	22	0.13
Fe	75.9	1.5	5.3	0.27	5.5	10	0.14
Zn	149	2.2	3.3	0.22	8.0	6.2	0.12
As	210	1.0	1.6	0.34	3.7	3.0	0.17
Se	233	0.9	1.3	0.34	3.3	2.6	0.18
Sr	321	3.2	1.8	0.18	11.7	3.5	0.096
Cd[d]	95.6	1.1	3.6	0.31	2.9	5.9	0.19
Sn	94.4	1.8	4.7	0.25	4.5	7.4	0.15
Ba	59.8	20.2	25	0.074	48.7	38	0.047
Pb[e]	110	2.3	4.5	0.215	8.8	8.9	0.111

[a]C(mdl) is in units of ng/cm^2. Data represents 100-second analysis for each of three secondary fluorescers.

[b]The elements aluminum through calcium in the periodic table are measured by their K_α X-rays using a titanium fluorescer.

[c]Elements titanium through strontium are measured by K_α X-rays using a molybdenum fluorescer.

[d]Elements zirconium through barium are measured by K_α X-rays using a samarium fluorescer.

[e]Heavy elements are measured with L_α or L_β X-rays using a molybdenum fluorescer.

Values of R_{bs} are tabulated for the following substrates: 5 mg/cm^2 Millipore esters of cellulose filter, 1 mg/cm^2 Nuclepore polycarbonate filter, 0.5 mg/cm^2 Mylar film, and ultrathin Formvar film. The minimum detectable limits in Table 5.1 are somewhat lower than the values in the report by Wagman et al.[2] and represent more recent measurements.[7]

The counting times were 100 seconds for each of three fluorescers for the case of the secondary fluorescer system and 100 seconds total for each of the other methods. The quantities C(mdl) and PBR(mdl) are derived from the sensitivity and background measurement using Equations 5.3 and 5.4 respectively. Values for C(mdl) are expressed in units of ng/cm^2.

Table 5.3   Duke PIXE System

Element	S	No Membrane			Formvar			Nuclepore			Millipore		
		R_{bo}	C(mdl)^a	PBR	R_{bs}	C(mdl)^a	PBR	R_{bs}	C(mdl)^a	PBR	R_{bs}	C(mdl)^a	PBR
Al													
Si													
S	1.9	0.018	23	2.5	0.5	123	0.46	90	1650	0.03	43	1100	0.05
K	260	0.018	0.17	2.5	0.9	1.2	0.35	120	14	0.03	460	27	0.02
V	530	0.020	0.09	2.3	0.3	0.34	0.60	50	4.4	0.05	180	8.4	0.02
Fe	410	0.025	0.13	2.1	0.14	0.30	0.88	6	1.9	0.13	22	3.7	0.07
Zn	190	0.020	0.25	2.3	0.025	0.28	2.1	0.3	1.0	0.6	1.6	2.2	0.25
As	90	0.018	0.50	2.5	0.025	0.6	2.1	0.06	0.9	1.3	0.18	1.6	0.80
Se	70	0.018	0.60	2.5	0.018	0.6	2.5	0.04	0.9	1.7	0.14	1.7	0.90
Sr	28	0.013	1.3	2.9	0.013	1.3	2.9	0.015	1.4	2.7	0.07	3.1	1.2
Cd	2.3	0.010	14	3.3	0.010	14	3.3	0.018	19	2.4	0.02	21	2.2
Ba													
Sn													
Pb^b	14	0.018	2.5	2.5	0.018	3.2	2.5	0.02	3.3	2.3	0.06	5.8	1.3

^{a}C(mdl) is units of ng/cm^2.   Data represents 100-second analysis at 3 MeV proton energy and 70 nA beam current.

^{b}The analysis of Pb is based on the Lβ line.   The Kα line is used for all other elements.

The results for these systems are also plotted in Figures 5.1, 5.2 and 5.3. Here the C(mdl) values (right-hand scale) for each system are compared to the relative abundances for aerosol pollutants as quoted by Cooper[8] and represent typical values and upper and lower levels for urban elemental concentrations in ng/m^3 (left-hand scale). The conversion factor from ng/m^3 to ng/cm^2 depends on the volume of air sampled per unit area of filter. For a typical Millipore membrane filter operating in an aerosol sampler, the upper limit before the filter becomes clogged is about 10 m^3/cm^2. In designing an aerosol sampler, one allows for variations in pollutant levels above this typical urban value, so we might expect that systems will operate with an allowable limit of 3 m^3/cm^2. In Figures 5.1, 5.2 and 5.3 the normalization chosen between the left- and right-hand scales is the somewhat conservative value of 1 m^3/cm^2.

One general comment about Figures 5.1, 5.2 and 5.3 is that each of the three systems is capable of determining the abundances of many elements present in typical aerosols. Certain elements such as selenium, cadmium and mercury are more difficult to detect except when present in elevated concentrations or when longer sampling times are used.

The normal filter media used in aerosol sampling would be represented by either the Nuclepore (1 mg/cm^2) or Millipore (5 mg/cm^2) substrates. The values of R_{bo} represent a lower limit of the method, assuming a minimum possible backing. The values quoted for Mylar and Formvar are included as representatives of detectability that can be achieved with other sample forms appropriate for the respective methods.

The wavelength dispersive system appears to be better for detecting elements ranging from fluorine to sulfur. The data shown in Figure 5.2 represents the C(mdl) for 28 elements, but the present EPA system can only measure 16 elements conveniently with the fixed crystals and would need to measure the others successively with the scanning crystal spectrometer. The wavelength dispersive method is capable of the best resolution, which reduces the number of cases where interelement interferences are important. If no crystals are used to measure the background levels, then it may be difficult to make accurate determinations of concentrations near the minimum detectable limit when the peak to background ratio is small.

The secondary fluorescer system with an energy dispersive Si(Li) detector can obtain information on about 45 elements in a 300-second analysis when all three fluorescers are used. According to Figure 5.2, about 25 elements from a typical urban atmosphere could be analyzed above the minimum detectable limits. An additional 20 elements would normally be reported as not detected but would be analyzed when a pollution episode causes their concentration to rise above the minimum detectable limit.

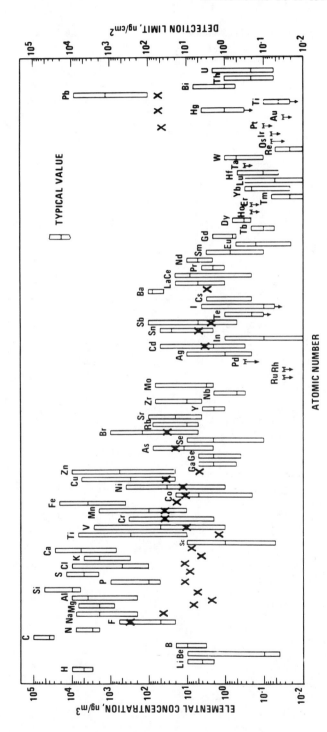

Figure 5.1 Minimum detectable limits in ng/cm² for the wavelength dispersive XRF system described in text are represented by the crosses. See text for discussion of conversion to ng/m³ for aerosol elemental concentrations according to right hand scale. The rectangles illustrate upper and lower ranges of concentrations for urban aerosols as reported in reference 8. Typical values are represented by the horizontal bars inside the rectangles.

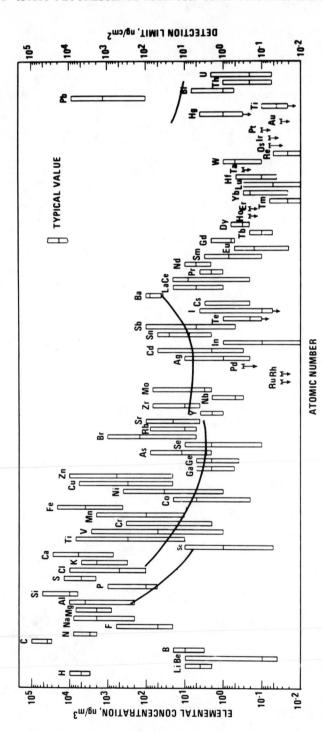

Figure 5.2 Minimum detectable limits in ng/cm² for the energy dispersive XRF system described in text. The curves are for the secondary fluorescers titanium, molybdenum and samarium (left to right) and for molybdenum (far right). See caption of Figure 5.1 for additional explanation.

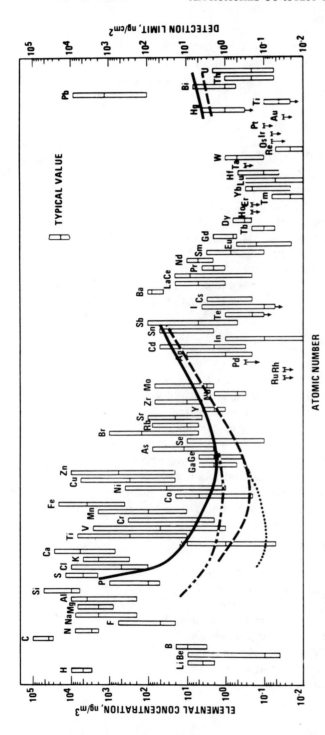

Figure 5.3 Minimum detectable limits in ng/cm^2 for the Duke and the FSU PIXE systems described in text. The curves illustrate the detection limits for the following situations: —— Millipore substrate (Duke),·—·— Nuclepore substrate (FSU), ——— Formvar substrate (Duke), ··· no substrate (Duke). See caption of Figure 5.1 for additional explanation.

According to Tables 5.1 and 5.2, the single element minimum detectable limits are comparable within a factor of 2 for elements with atomic numbers between 23 and 50. It should also be pointed out that in energy dispersive systems, the heavy elements from zinc to barium are measured using the high energy K_α radiation to avoid the interference in the L X-ray region that is discussed by Birks in Chapter 4 of this volume.

For the PIXE results in Table 5.3, the Duke system employed an absorber comprised of 0.13 mm of Mylar. The FSU results were obtained with a 0.7-mm Mylar absorber that had a hole amounting to 9% of the Si(Li) detector area.[5] This enhanced the sensitivity for light elements, as can be seen in Figure 5.3. The FSU results for the heavier elements are slightly worse than the Duke values. Both systems utilized about 70 nA of beam with an irradiation area of only about 0.5 cm^2. This small area is of practical significance because it allows one to employ miniature air samplers or air filter devices of small orifices as have been constructed at FSU.[5] The appropriate match between the area of irradiation and the sample deposit is essential if minimum detectabilities are to be achieved.

SUMMARY

On the basis of this comparison it is established that each of the three methods is capable of performing reasonable analytical measurement on air particulate samples. For each method, there are still possible improvements in techniques and instrumentation which could further enhance the capabilities of that technique. More detailed comparisons between the three XRF methods are difficult to make since additional factors such as accuracy, convenience, cost and reliability must enter into any practical considerations. The additional information presented in the accompanying chapters will contribute to an understanding of some of these factors.

ACKNOWLEDGMENTS

The authors would like to thank R. L. Bennett, K. T. Knapp and J. Wagman for kindly supplying the information for Table 5.1 of this chapter. This work was supported in part by the Environmental Protection Agency under Interagency Agreement with the U.S. Energy Research and Development Administration. Any conclusions or opinions expressed in this report represent solely those of the authors and not necessarily those of the Lawrence Berkeley Laboratory nor of the U.S. Energy Research and Development Administration.

REFERENCES

1. Currie, L. A. "Detection Power, Counting Statistics and the Real World," Chapter 25 of this volume.
2. Wagman, J., R. L. Bennett, and K. T. Knapp. "Simultaneous Multiwavelength Spectrometer for Rapid Elemental Analysis of Particulate Pollutants," Chapter 3 of this volume.
3. Jaklevic, J. M., B. W. Loo and F. S. Goulding. "Photon Induced X-Ray Fluorescence Analysis Using Energy Dispersive Detector and Dichotomous Sampler," Chapter 1 of this volume.
4. Jaklevic, J. M., F. S. Goulding, B. V. Jarrett and J. D. Meng. In *Analytical Methods Applied to Air Pollution Measurements*, R. K. Stevens and W. F. Herget, Eds. (Ann Arbor, Michigan: Ann Arbor Science Publishers, 1974), p. 123.
5. Nelson, J. W. "Proton Induced Aerosol Analyses: Methods and Samplers," Chapter 2 of this volume.
6. Walter, R. L., R. D. Willis, W. F. Gutknecht and J. M. Joyce. "Analysis of Biological, Clinical and Environmental Samples Using Proton-Induced X-Ray Emission," *Anal. Chem.* 46:843-855 (1974).
7. Knapp, K. T. and R. L. Bennett. Private communication.
8. Cooper, J. A. Battelle Northwest Laboratories Report BNWL-SA-4690.

6

USE OF X-RAY POWDER DIFFRACTION AND ENERGY DISPERSION ANALYSIS FOR THE INVESTIGATION OF AIRBORNE PARTICLES

R. L. Niemann and R. Jenkins

Philips Electronic Instruments, Inc.
Mount Vernon, New York

INTRODUCTION

Over the last 20 years, the X-ray powder diffractometer has proven to be an invaluable tool for both quantitative and qualitative phase analysis. Current technological advances and environmental problems are placing increasingly stringent requirements on the technique. These requirements are reflected in recent advances in instrumentation. During the 1960s, the diffractometer lagged far behind the spectrometer in terms of development and application. More recent work indicates that this trend is now being reversed. The introduction of high-powered diffraction tubes as well as the coupling of the small digital computer and the diffractometer with multispecimen handling facilities have done much to improve the sensitivity and speed of analysis. The advent of the energy dispersive spectrometer now provides the means of obtaining multielement analysis very rapidly, and many advantages of both the powder diffractometer (XRD) and the energy dispersive spectrometer can be retained by combining these two instruments. This represents a compromise between the speed of the energy dispersive spectrometer and the resolution of the conventional powder diffractometer. Thus, both phase and elemental information are obtained. In addition, the complexity of matching diffraction patterns from multiple phases with single standard patterns makes the availability of semiquantitative elemental information highly desirable.

As will be seen later, such information becomes extremely valuable because computerized search/match routines invariably give high priority to elemental data.

This chapter will consider some of the more recent work being done by a combination of X-ray powder diffractometers and energy dispersive spectrometers. Four broad areas will be considered: (1) typical applications of the powder diffractometer in the field of mixtures and particles, (2) combination of the energy dispersive diffraction and energy dispersive spectrometers in simultaneous diffraction and diffraction/spectrometry studies, (3) two combinations of the conventional energy dispersive spectrometer and the X-ray diffraction system, which provide unique new information for the diffractionist, and (4) use of combined data for successful qualitative identification of multiphase components.

AIRBORNE PARTICLE ANALYSIS BY CONVENTIONAL POWDER DIFFRACTOMETRY

The identification and quantitative determination of the crystalline constituents of mine and industrial dusts and the correlation of these with the appearance of silicosis in the lungs are among the useful and unique applications of X-ray diffraction analysis. The mineralogical composition, or state of chemical combination of the dust, has vastly greater pathological significance than the elemental composition deduced by chemical or X-ray fluorescence (XRF) analysis. Several applications have been described using the X-ray diffraction technique.[1,2]

Quantitative analysis of powder mixtures can be achieved following accurate measurements of intensities of powder diffraction lines. A good deal of research has been concerned with techniques for measuring intensities directly with the diffractometer. These have been primarily directed towards establishing relationships with intensities and integrated peak areas, as well as calibrations and absorption studies.[3] Since the early work of Sweany, Klaas and Clark,[4] and Clark and Reynolds,[5] the relationship between silicosis and α-quartz has been well known. Making analyses in the solid state is essential, since quartz as such is destroyed in a direct chemical method of analysis.

Complete dust analysis of large steel foundries correlates quartz content as measured by X-ray diffraction with lung radiographs for lesions and other nodules. Similar pathological lung conditions for workers with graphite, asbestos, talc, iron ores and beryllium compounds may also be correlated with X-ray diffraction studies. In all definite cases of silicosis, the crystalline quartz content in lung tissue varied from 0.2 to 2.5% (wt %). Since foundry workers had worked in the same dust-filled areas for many years, it was possible to study lung radiographs and correlate the active, arrested and absence of cases of silicosis. Quantitative analysis of industrial and mine dust by X-ray diffraction has been one of the more useful applications in the area of public health.

Bumsted later used X-ray diffractometry to study quartz content in coal dust. He found that the quartz content varied from 0.8 to 1.3% (wt %). Using the internal standard method and Millipore filter,[6] he identified some of the sample preparation and interference problems and established the analytical procedures for determination of permissible dust concentration.

It has been our experience that the typical detection levels that can be expected from conventional powder diffraction vary according to the matrix and sample nature, considering both powdered samples as well as those deposited on filter paper. The personal monitor provides a deposit on a 37-mm diameter filter. To accommodate such a filter, modifications have been made to sample holders as illustrated in Figure 6.1. This permits direct loading into a multiple cassette, enabling acquisition of data on the samples in the as-received condition. No chemical treatment or sample redeposition is necessary, and the results are comparable with those collected from prepared powders.

Figure 6.1   Field sampler (above) and modified Philips diffractometer sample holder (below). The field sampler accommodates 37-mm diameter filters, and the sample holder has a width of 35 mm.

Typical results on prepared mixtures are illustrated in Table 6.1. Conventional powder diffractometers have detection limits in the order of 0.10%. The data shown in the table are taken from a larger study involving prepared mixtures of both well crystallized and poorly crystallized materials in diluent matrices. Well defined and a more poorly defined crystalline matrices have been chosen in order to show this effect on the result.

Table 6.1 Typical Detection Limits for Diffraction Samples

Phase Diluent	Kaolinite Talc	α-SiO$_2$ Calcite	KBr Calcite	Calcium Oxalate Talc
Cts per %	5,300	7,100	7,150	2,500
Background	98,000	45,000	30,000	122,000
Lower limit of detection	0.12%	0.060%	0.048%	0.28%
Time	8 min	8 min	8 min	8 min
Blank counts	8,000	-570	-1570	2,764
Phase peak	24.40-25.34°	26.20-27.14°	26.64-27.58°	14.50-15.46°
Diluent peak	28.10-29.00°	28.98-29.90°	28.96-29.90°	28.10-29.00°
Div. slit	1°	1°	1°	1/2°
Rec. slit	1/2°	1/2°	1/2°	1/2°

CuKα 45kV 40mA   Long Fine Focus Tube   6° Take-off
P.C. Monochromator

From Table 6.1, it is apparent that one can expect in the order of 5000 counts/percent in a typical matrix. As the background rises due to the more amorphous diluents, the lower limit of detection is not as good, and a poorly crystallized diluent will produce results with as much as a factor of five worse detection limits than the crystalline matrices. Standard experimental conditions are given in the lower portion of Table 6.1, and these conditions are typical of most samples examined by X-ray diffraction. Lower limits of detection of the order of 0.10% are not as low as those experienced with the XRF spectrometer since modern XRF spectrometers have high power tubes, excellent vacuum and close target-sample coupling.

ENERGY DISPERSIVE DIFFRACTION/SPECTROSCOPY

Since energy dispersive diffractometry was introduced by Giessen and Gordon[7] and Buras et al.[8] in 1968, there has been considerable interest in this technique as applied to both the single crystal and powder methods. This chapter will concentrate only on those results obtained using the powder method.

Figure 6.2 shows the diffractograms obtained in our own laboratories,[9] using a specimen of α-quartz for both the energy dispersive and conventional methods. The lower diffractogram was obtained with a Si(Li) detector having a resolution of 175 eV and was acquired in 5 minutes. The upper diffractogram was obtained with a standard powder diffractometer in just less than 1 hour. Although the quality of the energy dispersive diffractogram is inferior to the classic case, the data acquisition time was an order of magnitude shorter. Probably the greatest advantage of the energy dispersive method is that of acquiring the whole diffraction pattern simultaneously. This has great potential when coupled with relatively short exposure times because it allows dynamic studies involving transient conditions such as phase changes due to changes in temperature or pressure.

Errors in the measurement of crystalline lattice spacings, d, due to misalignment can be quickly corrected in energy dispersive diffractometry. However, those errors due to displacement of the specimen or to beam penetration are more difficult to correct for than in the conventional diffractometer case. It will be appreciated that the higher energy photons penetrate much further into the specimen. Since in the energy dispersive diffractometry, d spacings are related by

$$d_{hkl} = 6.2/[E(keV) \cdot \sin\theta]$$

the higher energy photons correspond to small d values. In the conventional method, these small d values can be accurately measured; however, in energy dispersive diffractometry, errors in the energy due to penetration are reflected as errors in the small d spacings. Careful use of internal standards can do much to reduce this problem.

Voskamp[10] has published some interesting data using an energy dispersive diffractometer for the determination of retained austenite. Figure 6.3 compares the diffractograms of a 52100 steel obtained with energy dispersive and wavelength dispersive systems. From the diagrams, it is apparent that the resolution of the energy dispersive system is more than adequate to resolve the $(200)\alpha$, $(220)\gamma$, $(211)\alpha$, and $(311)\gamma$ lines, which are those commonly employed in the determination of retained austenite.

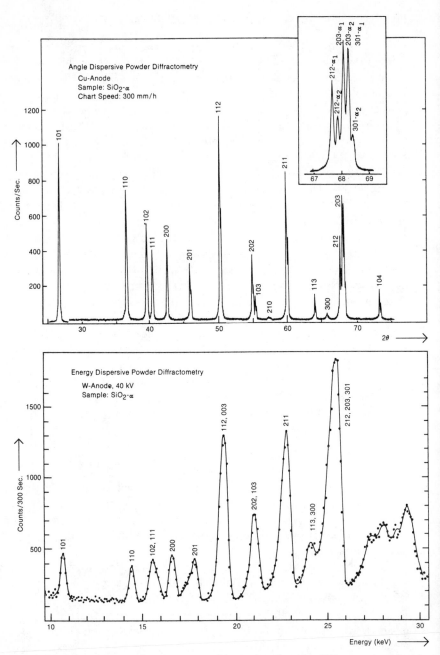

Figure 6.2 Diffractograms of alpha-quartz with angular dispersive diffractometer (above) and energy dispersive diffractometer (below).

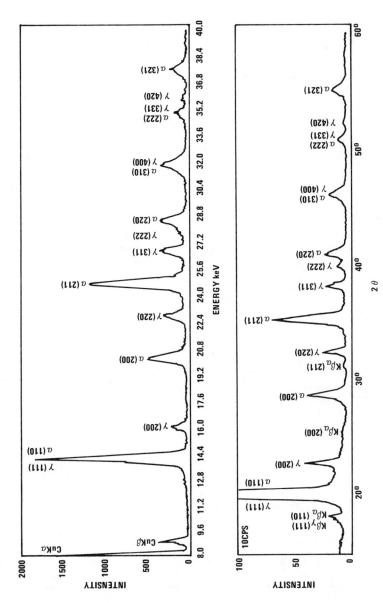

Figure 6.3 Comparison of diffractograms from retained austenite (18.3%) in 52100 steel. The upper trace represents energy dispersive detection with $2\theta = 25°$ for excitation using copper tube operated for 10 minutes at 50 kV and 24 mA. The lower trace represents conventional diffraction (Mo Kα/Zr filter, 50 kV 24 mA, 100 min).

Analysis times obtained with the energy dispersive technique were about two to three times shorter than with the conventional method, and Voskamp claims that it is possible to carry out the full quantitative analysis using the energy dispersive approach in about 6 minutes. Also the data acquisition cycle is considerably simplified.

Another area to be considered is the use of energy dispersive systems to give combined diffraction and spectrographic information. Diffracted photons and characteristic photons emitted by the specimen are readily discernable since the energy of the diffracted photons is angle-dependent. Figure 6.4 shows information taken from the work of Wen Lin[11] who has used the combined technique for the study of various alloys including this example, which is a copper, iron, nickel alloy. The upper diagram

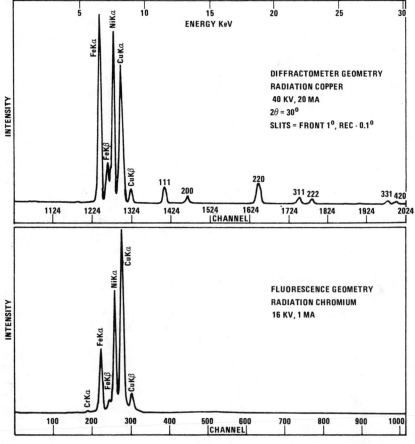

Figure 6.4   Diffractometer spectrum (top) and fluorescence spectrum (bottom) for Cu_4FeNi_3 alloy.

was taken with the detector angle set to satisfy the Bragg condition, and both diffraction and fluorescence emission lines were observed. The lower diagram was recorded with the detector angle deliberately misset from the Bragg condition. Only the fluorescence emission lines from the specimen are observed since their energies are independent of the detector angle.

ENERGY DISPERSIVE SPECTROMETRY WITH CONVENTIONAL X-RAY DIFFRACTION

Although energy dispersive diffraction may be useful in specific areas involving pure materials or simple mixtures of phases, in multiple phase analysis, the energy dispersive diffractometer invariably falls short of the conventional system in terms of adequate resolution. The best results seem to be obtained by equipping a conventional powder diffractometer, which obtains compound information, with an energy dispersive spectrometer, which obtains elemental information.[12] Such a combination retains the high resolution of the conventional diffractometer but still offers simultaneous acquisition of diffraction and spectrographic data. The schematic diagram shown in Figure 6.5 illustrates the overall system. The Si(Li) detector can be mounted opposite or at right angles to the diffractometer. The fluorescent emission from the specimen is collected by a tubular collimator mounted through the beam tunnel. Multispecimen handling with sample changer is possible along with the automation of the diffractometer. By use of an evacuated collimator, reasonable sensitivity can be obtained down to the element phosphorus ($Z = 15$). An added advantage of this set-up is that it is possible to isolate Compton-scattered primary radiation from the specimen. This yields valuable information because it can be used to estimate the amorphous content in a given diffraction specimen.[13] (J. J. Sahores pointed out that Compton-scattered intensity of the characteristic tube line can be used to compensate for variations of the specimen absorption as well as estimate the amorphous material. It is possible that the Compton-scattered intensity could be measured from the energy dispersive spectrum following some peak stripping procedure.)

We have recently started experimenting with a configuration different from that described in the previous paragraphs. Although this new configuration is not able to provide *simultaneous* diffraction and spectrographic information, it has some interesting possibilities. It has several features not available in the previously described system, such as a full vacuum path and markedly improved sensitivity when compared to the air or vacuum collimator model. In this configuration, the energy dispersive

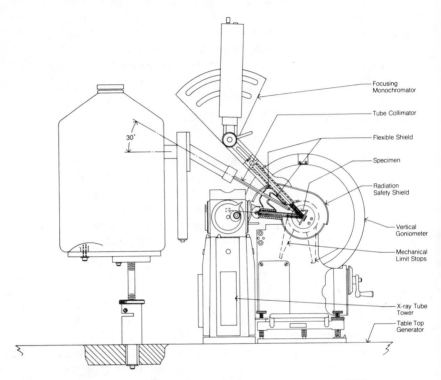

Figure 6.5   Schematic diagram of the energy dispersive spectrometer/diffractometer combination.

spectrometer chamber is interfaced to a free window of a diffraction tube tower, by means of a coupler called the Secondary Fluorescence Coupler (SFC). A variety of fluorescer/scatterer targets are provided and proper selection will cause a specimen to be irradiated either with the full spectrum from the diffraction tube, with almost monochromatic radiation of the operator's choice, or with the characteristic radiation from the diffraction tube. These targets, in combination with the variety of filters commonly fitted to a diffraction tube tower, offer a very wide range of excitation condition, giving excellent detection limits over a fairly wide atomic number range.

Although the sensitivity of the system for elemental analysis is somewhat less than that obtainable with a conventional XRF spectrometer, it must be remembered that time is not too critical since data acquisition time in routine powder diffractometry is of the order of 10 to 100 minutes.

Thus, a counting time of 10 minutes for the spectroscopic examination would not be considered unreasonable.

Figure 6.6 shows the SFC, which consists of a rotatable mount on which the fluorescer/scatterer can be mounted, the whole being located in a tubular shield. The fluorescer/scatterer itself can be rotated to optimize the incident and take-off angles. This device, as well as a diffractometer, can be connected to the X-ray ports of the diffraction tube tower.

Three fluorescer/scatterer combinations were used in this study, namely:

1. Lucite: a block of lucite 15 x 20 x 5 mm
2. Iron: a thin sheet of iron foil mounted on a curved aluminum backing (radius–100 mm)
3. Graphite: a pyrolytic graphite single crystal with a 2d value of 6.6Å.

The overall view of the assembly and experimental set-up is shown in Figure 6.7. The energy dispersive detector and sample presentation apparatus are part of a standard energy dispersive system. Samples are inserted into the chamber and the entire system evacuated. Various sizes of samples can be accommodated. For this operation, a special platform has been developed that will hold the same sample holders supplied with the powder diffractometer. Radiation from the SFC is directed upon the sample to be analyzed. A Si(Li) detector and multichannel analyzer are used to collect and display the radiation.

A series of measurements were made to determine the excitation efficiency of the iron fluorescer and the graphite and lucite scatterers. Twelve elements were chosen between Mg $(Z = 12)$ and Sb $(Z = 51)$, and absolute intensity data were obtained. Pure elements were chosen where these were available. Table 6.2 illustrates the actual count data obtained. The advantage of the iron secondary fluorescer for the lower atomic number elements is clearly seen, as is the superiority of lucite for the higher atomic numbers. Using the iron fluorescer, the determination of Mg is well within the capability of the system; the peak to background for this element is approximately 5:1 for pure magnesium. Some additional data are also included to illustrate some of the application possibilities and lower limits of detection obtained (Table 6.3). Some comparative data from a conventional X-ray spectrometer are also included.

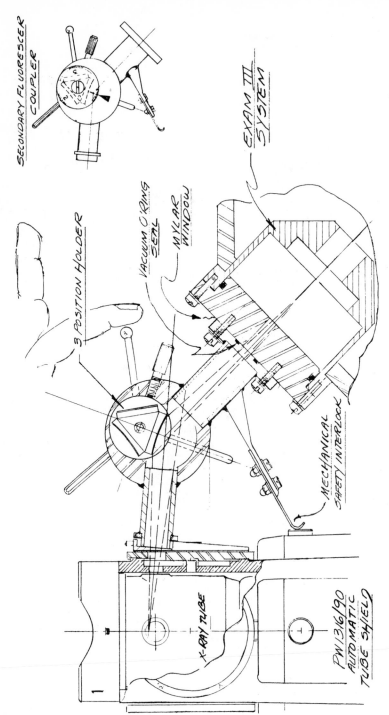

Figure 6.6 Schematic diagram showing relationship of diffraction tube, SFC and energy dispersive spectrometer (EXAM III).

Figure 6.7 Overall view of energy dispersive detector, sample measurement assembly, SFC, and diffraction tube tower for Philips Diffraction System.

Table 6.2 Typical Count Rates Obtained from Pure Elements

			Fluorescer/Scatterer	
Element	Energy, keV	Iron, Cts/Sec	Graphite, Cts/Sec	Lucite, Cts/Sec
Mg	1.25	2.0	0.4	1.0
Al	1.49	8.8	1.1	2.5
Si	1.74	7.4	1.1	1.8
P (ADP)[a]	2.02	15.9	3.1	6.5
S	2.31	19.3	4.3	8.8
Cl (NaCl)[a]	2.62	42.0	6.4	11.1
K (KCl)[a]	3.31	17.6	9.0	15.3
Ca (CaCO$_3$)[a]	3.69	76.2	11.7	20.2
Fe	6.40	30.3	101.9	234.2
Cu	8.04	29.0	135.3	185.1
Br (NaBr)[a]	11.92	60.5	176.8	310.1
Sb	26.36	64.2	93.5	138.2

[a]Normalized to 100%.

Table 6.3 Lower Limits of Detection from the Secondary Fluorescence Coupler
and Crystal Spectrometer

Matrix	Element	SFC	Crystal Spectrometer
Low alloy steel	Molybdenum	0.04%	0.0008%
Brass	Zinc	0.15%	0.0014%
Low alloy steel	Iron	0.005%	–
Low alloy steel	Nickel	0.26%	0.0024%
Low alloy steel	Manganese	0.97%	0.0034%
Low alloy steel	Chromium	0.10%	0.0010%
Low alloy steel	Vanadium	0.17%	0.0002%
Fertilizer	Potassium	0.75%	–
Fertilizer	Chlorine	1.1%	–
Fertilizer	Sulfur	1.5%	0.0016%

COMBINATION OF DATA IN MULTIPLE-PHASE IDENTIFICATION

The ability to obtain elemental as well as phase information can be a powerful tool to the diffractionist. The number of standard patterns available from the Joint Committee on Powder Diffraction Standards (JCPDS) now exceeds 26,000. The number of possible solutions from a search can be greatly reduced if the search is aided with information additional to the "d-spacings" and intensities. The introduction of computerized search/match routines has provided the user with a rapid and systematic means of searching a large number of possibilities. However, in multiple-phase mixtures, the large amount of data may yield a significant number of possible matches. Typically, we have observed up to 200 possible compounds from a given search/match program. However, if the user can also introduce specialized information into the search/match, the solution will be quicker and more straight forward. The more successful search/match routines give a high priority to elemental data. This point is best illustrated with the data shown in Table 6.4. In a mixture of $BaCl_2$, KI, and binder, an attempt was first made to identify the phases from the diffraction information only. The number of possible solutions was very large, and the output indicated that the compounds $BaCl_2$ and KI were 34th and 86th of more than 86 choices. However, with the addition of elemental information collected by the energy dispersive spectrometer, the number of possibilities dropped dramatically. The combination of both types of data rapidly and correctly sorted out the complex

Table 6.4   Effect on Ranking of Elemental Data in Computer Search Matching

Conditions	Number of Possibilities	Rank	
		BaCl₂	KI
No elemental data given	> 86	34	86
Positive elemental data given (i.e., Ba, I, K, Cl)	26	5	16
Positive and negative elemental data given, i.e., elements present (E.D.S.): Ba, I, K, Cl. Additional elements not present: Z > 15. Possible element: Z ≤ 15	5	1	3

mixtures of multiphase unknowns. $BaCl_2$ and KI were ranked in the top five positions in order of probability.

In the analysis of airborne particles the identification of phases from diffraction data alone presents particular problems because of the limited quantity of material, the relatively high background from the substrate support material and the rather heterogeneous nature of the deposits. Although we have been successful in obtaining reasonable diffraction patterns from as little as 0.1 mg/cm² of material, complete phase identification is almost impossible without the benefit of a reasonably good semiqualitative elemental analysis. To obtain such an analysis with an off-line XRF spectrometer may also present problems due to the need to remount the sample. An instrument combination of energy dispersive spectrometry and SFC along with powder diffraction is particularly useful since it does allow the acquisition of diffraction and spectrographic data on one and the same specimen.

REFERENCES

1. Klug, H. P. and L. E. Alexander. *X-Ray Diffraction Procedures for Polycrystalline and Amorphous Materials* (New York: J. Wiley, 1954), pp. 410-433.
2. Loranger, W. F. and G. L. Clark. *Anal. Chem.* 26:1413 (1954).
3. Jenkins, R. and J. L. deVries. *X-Ray Powder Diffractometry* (Eindhoven: N. V. Philips Gloeilampenfabrieken, 1970).
4. Sweany, H. C., R. Klaas and G. L. Clark. *Radiology* 31:299 (1938).

5. Clark, G. L. and D. H. Reynold. "Quantitative Analysis of Mine Dusts: An X-Ray Diffraction Method," *Ind. Eng. Chem. Anal. Ed.* 8:36 (1936).
6. Bumsted, H. E. *Amer. Indust. Hygiene Assoc. J.*   150-158 (April 1973).
7. Giessen, B. C. and C. E. Gordon. *Science* 159 (1968).
8. Bevas, B. *et al.* *Institute of Nuclear Research Report* 894/11/PS (Warsaw, 1968).
9. Walinga, J.   Philips Analytical Equipment Bulletin (November 1972).
10. Voskamp, A. P. *Adv. X-Ray Anal.* 17:124 (1973).
11. Lin, Wen. *Adv. X-Ray Anal.* 16:298 (1972).
12. Jenkins, R. *Norelco Reporter* 20(3):22-30 (1973).
13. Sahores, J. J. *Adv. X-Ray Anal.* 16:186 (1972).

SECTION II

SAMPLE COLLECTION TECHNIQUES
FOR AEROSOL ANALYSIS

7

APPLICATION OF THE DICHOTOMOUS SAMPLER TO THE CHARACTERIZATION OF AMBIENT AEROSOLS

Thomas G. Dzubay and Robert K. Stevens

Environmental Sciences Research Laboratory
U.S. Environmental Protection Agency
Research Triangle Park, North Carolina

C. M. Peterson

Environmental Research Corporation
St. Paul, Minnesota

INTRODUCTION

In 1973, a program was undertaken to develop a dichotomous sampler capable of separately collecting the respirable and nonrespirable fractions of atmospheric aerosols for subsequent gravimetric, chemical and X-ray fluorescence (XRF) analyses. This was needed as an alternative to the high-volume sampler, which collects particles in a single size range and is not designed to accommodate the membrane filters that are required for XRF analysis. However, for the dichotomous sampler to be compatible with the high-volume sampler, it must be possible to determine the total mass of the collected aerosol.

In the development of the dichotomous sampler, emphasis was placed upon achieving a design that yields high sampling accuracy. Conventional impactors were not used for the size separation because of known particle bounce errors associated with the impaction surfaces.[1,2] If any type of dry impaction surface is used, a significant fraction of the larger particles can be expected to bounce from the surface and be lost on the impactor walls or be collected as a contamination among the fine particles on subsequent stages.[1,2] Although the impaction surface can be coated with a

95

layer of grease to prevent particle bounce errors,[1,2] this approach was not used in the present development effort because the grease could be an interference for some of the required analyses.

The technique of virtual impaction was selected to perform the aerodynamic separation between the respirable and nonrespirable fractions.[3-5] In the virtual impactor, particles are impacted into a void where they are collected by filtration.[3] The use of this technique eliminates the impaction surface that must be coated with grease in ordinary impactor devices. By using Teflon*membrane filters, the possibility of forming artifacts (such as SO_2 → sulfate conversion) on the collection surfaces is minimized.

A prototype dichotomous sampler that utilizes virtual impaction was evaluated in the field in St. Louis during the summer of 1973.[3] The evaluation demonstrated that the dichotomous sampler is compatible with the requirements of both gravimetric mass analysis and X-ray fluorescence elemental analysis. More recently, a network of ten automated dichotomous samplers has gone into operation as a part of the Regional Air Pollution Study in St. Louis.[4] This chapter describes a manual dichotomous sampler designed for routine field use.

DESCRIPTION

Figure 7.1 shows photographs of a manual dichotomous sampler consisting of an aerosol inlet, a virtual impactor particle separator, vacuum pumps, and flow servocontroller. Also shown at the left in the photograph is the filter holder assembly and a pair of filters used to collect fine and coarse particles from the atmosphere.

Aerosol Inlet

The aerosol inlet must quantitatively transport particles into the dichotomous sampler without transporting raindrops or insects. This becomes increasingly difficult as the particle diameter becomes larger than 10 μm. Inertial effects and gravitational settling make it difficult to transport quantitatively larger particles into the inlet as the wind speed varies. Therefore, it was decided to attempt to collect particles only up to 20 μm in diameter and to reject larger ones, which often originate as windblown dust rather than from combustion sources.

Figure 7.2 shows a prototype aerosol inlet that was designed for this purpose. Air is drawn into the inlet at a relatively high flow rate (233 liters/min) for adequate response to particles up to 20 μm in diameter. Particles larger than 20 μm are removed by an annular-shaped virtual impactor that operates on the same principle as the virtual impactor particle

---

*Registered trademark of E. I. duPont de Nemours & Company, Inc., Wilmington, Delaware.

Figure 7.1 Manual dichotomous sampler. The close-up view shows the filter holder assembly with coarse and fine particles collected at the left and right, respectively.

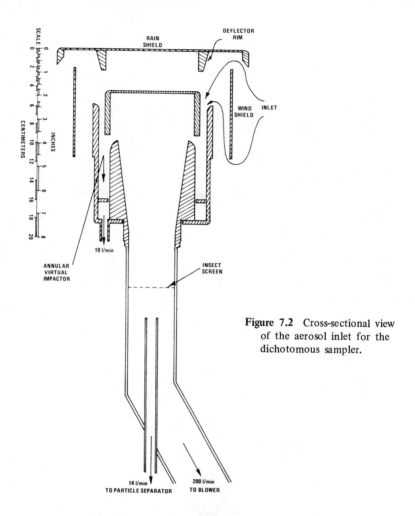

Figure 7.2 Cross-sectional view of the aerosol inlet for the dichotomous sampler.

separator described below.   The aerosol inlet has a deflector rim on the bottom of the rain shield to direct particles into the sampler during conditions of high wind speed.   At the bottom of the inlet, particles are isokinetically sampled at a rate of 14 liters/min from the air stream and transported to the dichotomous sampler.   The flow of the remaining air stream is maintained by a blower type of pump.

Virtual Impactor Particle Separator

Figure 7.3 shows a cross-sectional view of the particle separator used to deposit fine and coarse particles on membrane filters.   For this device,

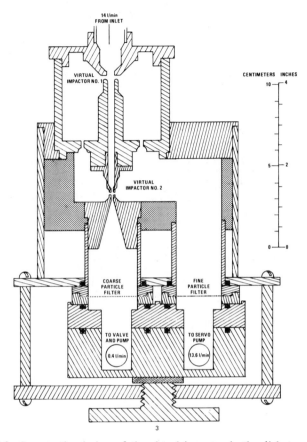

Figure 7.3 Cross-sectional view of the virtual impactor in the dichotomous sampler.

the 50% cutpoint separation diameter for fine and coarse particles of unit density occurs at 3.5 μm. Two virtual impactor stages are operated in series.[4]   A servosystem and carbon vane pump are used to maintain a flow rate of 13.6 liters/min through the fine particle filter; a needle valve is used to adjust the coarse particle flow rate to 0.4 liter/min. Because there is a small flow of air through the coarse particle filter, a proportionally small amount of fine particle mass is collected on the coarse particle filter. To correct for this effect, particle concentrations in the two size ranges are determined from the equations:

$$C_f = M_f/(tF_f)$$
$$C_c = [M_c - (M_f F_c/F_f)] / [t(F_f + F_c)].$$

where C_f and C_c =   the atmospheric concentrations of the fine and coarse
                                      particle fractions, respectively in $\mu g/m^3$

        M_f and M_c =   the masses collected on the fine and coarse particle
                                        filters, respectively in μg

         F_f and F_c =   the flow rates through the fine and coarse particle
                                        filters, respectively in m^3/min

        t   =   the sampling time in minutes.

Figure 7.4 shows the fractionation characteristics and loss curve for the particle separator pictured in Figure 7.3. For this particle separator, a concentration factor of $F_f/F_c = 34$ is used. Attempts to increase the concentration factor by lowering the value of F_c have been accompanied by an increase in particle losses in the virtual impactor stages.[4] Evaluation of the design shown in Figure 7.3 indicates that the particle deposits are uniform across the filters.

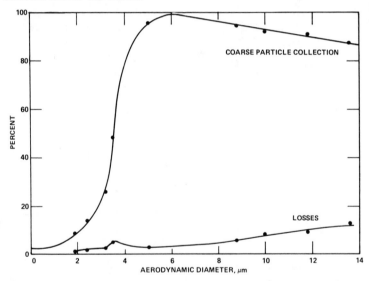

Figure 7.4   Fractionation characteristics and loss curves for the virtual impactor
measured using solid uranine particles.

The losses plotted in Figure 7.4 were measured using solid particles of uranine produced with a vibrating orifice aerosol generator.[6] The losses were between 1 and 13% for solid particles in the 2 to 14 μm range. According to Loo, the losses for liquid particles can be somewhat larger than the losses for solid particles.[4] It is believed that the losses for both solid and liquid particles are negligible for diameters below 2 μm. An additional small correction of a few percent will have to be applied to account for the coarse particle losses. However, studies of the contours of the jets of the virtual impactor are being made to further reduce the losses.

Flow Rate Controller

The membrane filters, which are required for XRF analysis, have a significant flow resistance,[7] and this flow resistance increases as the filter becomes loaded with particles. According to Equations 7.1 and 7.2, any errors or drift in the fine particle flow rate F_f would cause serious errors in the calculated concentration. Hence, it is necessary to use a servo-mechanism to control the flow rate through the fine particle filter to an accuracy of 1%. Flow rate regulators are available that operate on the following principles: (1) anemometer sensor and variable power to pump motor, (2) anemometer sensor and variable orifice between filter and pump, and (3) pressure differential flow sensor and motor-controlled needle valve between filter and pump. A program is underway to evaluate the reliability and performance of the above three methods of flow control. The flow rate through the coarse particle filter is less critical; controlling F_c to an accuracy of 10% may be sufficient.

EXPERIMENTAL RESULTS

Eight dichotomous samplers were performance-tested beside high–volume samplers for 20 days during August and September of 1975 at a rural and at an urban site in the vicinity of St. Louis, Missouri. The earlier version of the sampler was identical to the most recent design shown in Figures 7.1, 7.2 and 7.3 except that the coarse particle flow rate F_c was 0.28 liter/min rather than the present 0.4 liter/min, and the distance between the coarse particle filter and the last jet of the second virtual impactor was about half that shown in Figure 7.3. Fluoropore* filters with 1-μm pore diameters were used for particle collection. According to the measurements by Liu,[7] the efficiency is greater than 99.9% for particles with diameters above 0.034 μm.

To determine the collected mass, the Fluoropore filters used in the dichotomous samplers were weighed to a precision of 10 μg using an electrobalance in a room with the relative humidity adjusted to 40%. Immediately before the filters were weighed, they were passed in front

---

*Mention of commercial products or company names does not constitute endorsement by the U.S. Environmental Protection Agency.

of a ^{210}Po radioactive source to remove any electrostatic charge. The glass fiber filters used in the high–volume samplers were weighed using a mechanical balance.

The Fluoropore filters were subsequently analyzed using an energy dispersive X-ray fluorescence spectrometer,[3,8] which was calibrated using thin vapor-deposited films made by Micro Matter Co., Seattle, Washington. The X-ray spectra were analyzed using a stripping procedure described by Jaklevic et al.[8]

Figure 7.1 shows a photograph of typical filters used in the dichotomous sampler. The fine particle deposit is black; the coarse particle deposit is a light tan. Table 7.1 shows the results of the XRF and gravimetric analyses of the fine and coarse particles collected at the urban and rural sites. These data indicate that most of the sulfur, bromine, and lead from combustion sources occurs in the fine particle fraction, and most of the aluminum, silicon, calcium, titanium and iron occurs in the coarse particle fraction. The sulfur constitutes about 12.5% of the mass of the fine particle fraction. If one assumes that the sulfur is in the form of ammonium sulfate,[3] then this accounts for about half of the mass in the fine particle fraction.

Table 7.1 Mass and Percentage Composition of Size Fractionated St. Louis Aerosol Averaged over the Period from August 18 to September 7, 1975

	Urban[a]		Rural[b]	
	Fine (%) 29 μg/m^3	Coarse (%) 22 μg/m^3	Fine (%) 26 μg/m^3	Coarse (%) 15 μg/m^3
Si	1	8	0.5	4
S	12.5	1.4	12.6	0.9
K	0.4	1.2	0.3	0.9
Ca	0.7	8.2	0.5	4.2
Ti	1.1	2.0	< 0.1	0.2
Fe	1.4	4.8	0.3	1.3
Zn	0.35	0.20	0.13	0.15
Br	0.33	0.16	0.06	0.04
Pb	2.2	0.60	0.51	0.11

[a]Located at the Missouri Botanical Garden in St. Louis.
[b]Located in an agricultural area in Illinois, 40 km south of St. Louis.

Table 7.2 shows the standard deviations for mass, sulfur, lead and calcium determined for four samplers operated simultaneously at each site. For sulfur, lead and total mass, the standard deviation in the fine particle measurements is quite small. However, for the quantities measured in the

Table 7.2   Comparability of Four Dichotomous Samplers Run Simultaneously
at the Urban Site of Table 7.1

	Fine Particles		Coarse Particles	
	σ $(\mu g/m^3)^a$	σ $(\%)^a$	σ $(\mu g/m^3)^a$	σ $(\%)^a$
Mass	2	7	4	18
S	0.2	5	0.06	19
Ca	0.06	30	0.5	29
Pb	0.04	6	0.03	26

[a]Standard deviation σ expressed in $\mu g/m^3$ and as a percentage of the average concentration.

the coarse particle range, the standard deviation is somewhat large. Visual inspection of the coarse particle filters revealed that the deposits were sometimes not uniform but were concentrated in a 1-cm-diameter spot at the center of the filter. Subsequent laboratory investigation revealed that the nonuniformity can be eliminated by increasing the coarse particle flow rate F_c and by increasing the distance between the virtual impactor jet and the coarse particle filter. These observations led to the improved design shown in Figure 7.3, which does produce uniform deposits on both filters.

Figure 7.5 shows a comparison of the total (fine plus coarse) mass concentrations determined from the dichotomous and high–volume samplers. The excellent agreement indicates the potential of the dichotomous sampler to be used as a substitute for the high–volume sampler. Further studies are required in a variety of geographical areas to determine the effect of particles larger than 20 μm on such comparisons.

CONCLUSION

The dichotomous sampler is well suited to the collection of atmospheric particles on membrane filters for subsequent XRF and gravimetric analysis. Present results indicate that the mass concentrations measured using the high–volume and dichotomous samplers are equivalent. This may not be the case when there are high winds that cause particles larger than 20 μm in diameter to be suspended in the atmosphere. However, such particles are not necessarily of interest in air pollution monitoring. The dichotomous sampler offers the advantage of separately collecting the respirable and nonrespirable fractions. The virtual impactor particle separator can produce

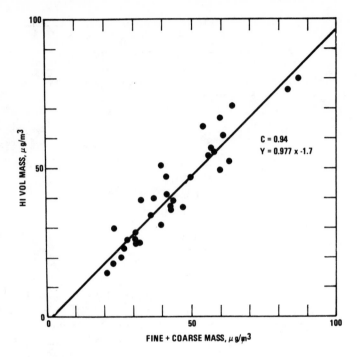

Figure 7.5   A comparison of the total mass concentrations determined using high–volume and dichotomous samplers.

uniform deposits of both fine and coarse particles that are suitable for XRF analysis. The respirable fraction is observed to contain most of the total lead and sulfur, which are assumed to originate from combustion sources.

REFERENCES

1. Wesolowski, J. J., W. John, W. Devor, T. A. Cahill, P. J. Feeney, G. Wolfe and R. Flocchini. "Collection Surfaces of Cascade Impactors," Chapter 9 of this volume.
2. Dzubay, T. G., L. E. Hines and R. K. Stevens. "Particle Bounce Errors in Cascade Impactors," *Atmos. Environ.* 10:229-234 (1976).
3. Dzubay, T. G. and R. K. Stevens. "Ambient Air Analysis with Dichotomous Sampler and X-Ray Fluorescence Spectrometer," *Environ. Sci. Technol.* 9:663-668 (1975).
4. Loo, B. W., J. M. Jaklevic and F. S. Goulding. "Dichotomous Virtual Impactors for Large Scale Monitoring of Airborne Particulate

Matter," in *Fine Particles*, B. Y. H. Liu, Ed. (New York: Academic Press, 1976), pp. 311-350.

5. Stevens, R. K. and T. G. Dzubay. "Recent Development in Air Particulate Monitoring," *IEEE Trans. on Nucl. Sci.* NS-22:849-855 (1975).

6. Berglund, R. N. and B. Y. H. Liu. "Generation of Monodisperse Aerosol Standards," *Environ. Sci. Technol.* 7:147-153 (1973).

7. Liu, B. Y. H. and G. A. Kuhlmey. "Efficiencies of Air Sampling Media," Chapter 8 of this volume.

8. Jaklevic, J. M., F. S. Goulding, B. V. Jarrett and J. D. Meng. "Applications of X-Ray Fluorescence Techniques to Measure Elemental Composition of Particles in the Atmosphere," in *Analytical Methods Applied to Air Pollution Measurement*, R. K. Stevens and W. F. Herget, Eds. (Ann Arbor, Michigan: Ann Arbor Science Publishers, 1974), pp. 123-146.

8

EFFICIENCY OF AIR SAMPLING FILTER MEDIA

Benjamin Y. H. Liu and Gregory A. Kuhlmey

Particle Technology Laboratory
Mechanical Engineering Department
University of Minnesota
Minneapolis, Minnesota

INTRODUCTION

Filtration is a widely used technique for sampling atmospheric aerosols. With the use of a suitable filter, airborne particles can be collected for subsequent analysis by physical or chemical means, such as X-ray fluorescence. To be usable in a given application, a filter must have physical or chemical properties compatible with the available analysis technique, including a low background in the material to be analyzed. In addition, the filter efficiency and flow resistance must also be known.

The efficiency of a filter for atmospheric aerosols can be determined by sampling the aerosol with two or more filters in series and analyzing the collected particles with the available analysis technique. However, since the filter efficiency is dependent upon the size distribution of the aerosol, such measurements are generally valid only for the specific aerosol used, and the results cannot be easily generalized to aerosols with different size distributions.

The efficiency of a filter can also be measured with laboratory-generated monodisperse aerosols. For instance, the conventional DOP test[1] uses monodisperse aerosols of DOP (dioctyl phthalate), a nontoxic oily substance of low vapor pressure. The aerosols are generated by vaporization and condensation, and their penetration through the filter is measured with a light-scattering photometer. The test is quite convenient and is capable of giving quick measurement results. In addition, it is also highly sensitive. Under optimal conditions, filter penetrations as small as 1 part in 10^6 can be measured. However, due to insufficient light scattering from small particles, the method is limited to particles larger than 0.2 μm in diameter.

In applying the conventional DOP test, 0.3-μm diameter particles are generally used. This is done for two reasons. First, the 0.3-μm particle size represents the smallest practical size for use with the light-scattering photometer. Second, according to the classical filtration theory, the efficiency of a fibrous filter at low-to-moderate pressure drops is a minimum in the vicinity of 0.3 μm because in the absence of electrostatic effects particles are captured in a filter primarily by the mechanisms of diffusion, inertial impaction and interception. Both larger and smaller particles are removed with greater efficiencies than particles in the 0.3-μm size range because of the higher diffusion coefficient of the smaller particles on the one hand and the greater inertial and interception effects of the larger particles on the other hand. In general, the efficiency-versus-particle-size curve would appear as in Figure 8.1 with a minimum in the vicinity of 0.3 μm. Thus, the 0.3-μm DOP test has come to be regarded as the most stringent test for filters. It is widely used for testing high perform-ance filters such as those used in protective face masks, clean rooms, and nuclear power plants. In addition, it has also been used by some authors[2] for testing filters used in air sampling.

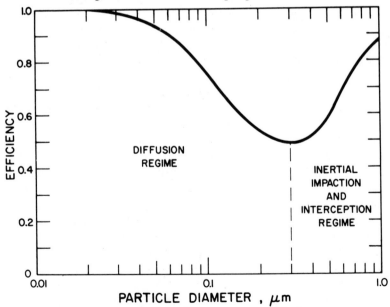

Figure 8.1   A typical filter efficiency-versus-particle-size curve.

Although the classical filtration theory predicts that fibrous filters operating at low-to-moderate pressure drops will have a minimum in their efficiency curves in the vicinity of 0.3 μm, recent theoretical[3] and experimental[4] studies have shown that for membrane and Nuclepore filters operating at moderate-to-high pressure drops, the minimum efficiency tends to occur at much smaller particle sizes. For these filters, which are widely used in X-ray fluorescence applications, the classical DOP test is not a valid test for the minimum filter efficiencies.

In this chapter, a new experimental technique is described that is capable of generating the entire efficiency-versus-particle-size curve similar to that shown in Figure 8.1. At the present time, the method is limited to particles in the 0.034-1-μm diameter range. However, with further improvement in the apparatus, this range can be extended if necessary.

EXPERIMENTAL

The experimental system used here has been described in some detail in a previous publication.[4] However, the apparatus and procedure used are briefly reviewed in order to facilitate understanding of the results.

Figure 8.2 is a schematic diagram of the filter testing apparatus used.

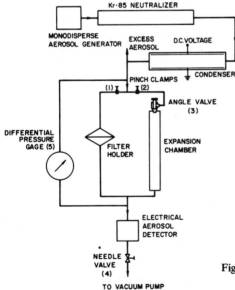

Figure 8.2 Filter testing apparatus making use of the electrical aerosol detector.

The principal components include a monodisperse aerosol generator, a krypton-85 aerosol charge neutralizer, and an electrical aerosol detector. The aerosol generator used is of the atomization-condensation type described by Liu and Lee.[5] The aerosols used were DOP particles with median particle diameters ranging from 0.034 to 1 μm and geometrical standard deviations of 1.4 to 1.2. When the filter efficiency was measured, the aerosol was allowed to pass through the filter holder containing the filter under test and also through a parallel chamber with an expansion valve. By measuring the aerosol concentration downstream of the filter holder and downstream of the expansion chamber, the relative penetration of the aerosol through the filter and through the expansion chamber was determined. After correcting for the particle loss in the expansion chamber, which was usually quite small, the filter efficiency was determined. The pressure drop across the filter was varied in these experiments between 1 and 30 cm Hg. By varying the size of the particles produced by the monodisperse aerosol generator, the filter efficiency could be determined as a function of particle size for each of several preselected pressure drops across the filter medium.

In these experiments, the flow resistance of the filter media was also measured. The apparatus used is shown in Figure 8.3. The volumetric

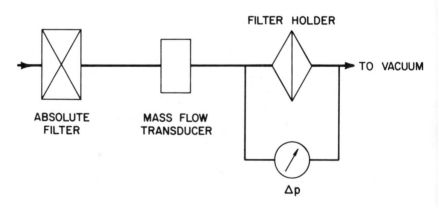

Figure 8.3 Apparatus for measuring flow resistance of filter media.

flow rate of filtered air at the inlet to the filter was measured as the pressure downstream of the filter was varied by means of a needle valve. The flow rate was measured with a calibrated mass flow transducer. Based on the known flow area of the filter, the inlet face velocity was then computed as a function of the pressure drop across the filter. It was found that the supporting screen in the filter holder contributed negligibly to the total pressure drop across the system; consequently, no correction was made for the pressure drop caused by the filter screen support.

DESCRIPTION OF FILTERS TESTED

A total of 13 filters have been tested so far. With the exception of two filters, all are standard filter media available from commercial sources. These include the Nuclepore filters (Nuclepore Corp., 7035 Commerce Circle, Pleasanton, California 94566) of 0.4 to 8 μm pore size, the Teflon membrane filters manufactured by the Millipore Corp. (Bedford, Massachusetts 01730) and marketed under the trade names of Mitex and Fluoropore Filters, and the Tissue Quartz filters manufactured by the Pallflex Products Corp. (Kennedy Drive, Putnam, Connecticut 06260). The two filters tested that were not standard commercial filters are the 1.2-μm cellulose acetate membrane filter, manufactured specially by the Nuclepore Corp. for use by EPA in their dichotomous atmospheric aerosol samplers,[6,7] and a 1-μm pore Kinar filter being developed by the Millipore Corp. and supplied to EPA for evaluation purposes. The Nuclepore filters have been found by a number of investigators to be suitable for X-ray fluorescence work. The cellulose membrane filters manufactured by the Nuclepore Corp. are being used for the elemental analysis of deposited particles by X-ray fluorescence. Both the Teflon membrane and Kinar filters are potential candidates for X-ray fluorescence applications. While the usefulness of the Pallflex Tissue Quartz filter for X-ray fluorescence is not known, it has been used and found satisfactory for mass concentration measurement by beta attenuation and for chemical analysis by extractive chemical methods.

RESULTS

Some typical experimental results for several of the filters tested are shown in Tables 8.1 through 8.5. For each of the filters tested, we give the inlet face velocity as a function of the pressure drop across the filter, and the filter efficiency at each of several preselected particle sizes and filter pressure drops. The efficiency values shown include an estimate of the experimental uncertainties. The uncertainty is seen to vary from about ±0.01% for large particles, which tend to give a high electrical aerosol detector output, to as large as ±1% for small particles, which tend to produce a small detector output.

Table 8.1   Flow Resistance Measurements for Various Filter Media

Pressure drop ΔP, cm Hg	Velocity v_0, cm/sec			
	0.4 μm Nuclepore (polycarbonate)	1.2 μm Cellulosic Nuclepore (esters of cellulose)	Tissue Quartz Pallflex	1.0 μm Millipore Kinar (min-max)[a]
1.0	4.4		21.0	11.7-16.4
1.5	6.6		32.7	17.2-24.9
2.0	9.0		44.4	22.2-32.0
3.0	13.6		65.5	32.7-49.1
5.0	22.2		107.6	55.3-81.0
7.0	31.4		146.5	76.4-113.0
10.0	43.3		205.8	108.0-154.0
10.8	—	77.5		
15.0	59.3			160.0-228.0
20.0	65.5			$>$ 206.0
30.0	109.0			

[a]Minimum (membrane #1) and maximum (membrane #3) for three membranes tested.

Table 8.2   Filter Efficiency versus Pressure Drop for 0.4-μm Nuclepore Filter

ΔP, cm Hg	Percent Efficiency						
	D_p = 0.034	0.05	0.07	0.10	0.15	0.20	0.30 μm
1	99.3 ±0.4	99.2 ±0.4	99.3 ±0.3	99.0 ±0.2	98.9 ±0.2	99.4 ±0.1	99.92 ±0.08
3	96.1 ±0.3	96.6 ±0.3	93.4 ±0.1	93.4 ±0.1	95.50 ±0.06	97.79 ±0.04	99.42 ±0.03
10	89.1 ±0.5	90.3 ±0.6	87.1 ±0.2	89.7 ±0.2	95.04 ±0.08	97.61 ±0.06	98.64 ±0.05
30	80.6 ±0.3	82.4 ±0.3	87.6 ±0.1	91.97 ±0.08	97.03 ±0.05	98.55 ±0.03	99.22 ±0.03

Table 8.3 Filter Efficiency versus Pressure Drop for 1.2-μm Pore Cellulose Acetate Membrane Filter

ΔP, cm Hg $D_p =$	0.034	0.05	0.07	0.10	0.15	0.20	0.30	0.50	0.70	1.0 μm
					Percent Efficiency					
10.8	> 99.6	99.8 ±0.2	99.84 ±0.08	99.88 ±0.06	99.92 ±0.04	99.94 ±0.03	99.94 ±0.02	99.93 ±0.03	99.95 ±0.01	99.97 ±0.01
10.8[a]	99.8 ±0.2	99.7 ±0.2	99.84 ±0.08	99.89 ±0.06	99.89 ±0.03	99.92 ±0.02	99.98 ±0.02	99.98 ±0.02	99.98 ±0.01	99.98 ±0.01

[a]Second filter tested under identical conditions.

Table 8.4 Filter Efficiency versus Pressure Drop for Pallflex Tissue Quartz Filter

ΔP, cm Hg $D_p =$	0.034	0.05	0.07	0.10	0.15	0.20	0.30	0.50	0.70	1.0 μm
					Percent Efficiency					
1	94.0 ±1.0	> 99.3	99.8 ±0.2	99.7 ±0.2	99.84 ±0.08	99.94 ±0.06	> 99.95	> 99.96	> 99.97	> 99.96
3	> 99.4	99.3 ±0.2	99.48 ±0.09	99.20 ±0.06	99.56 ±0.04	99.82 ±0.03	99.94 ±0.02	99.94 ±0.02	99.93 ±0.02	99.96 ±0.02
10	98.1 ±0.5	98.1 ±0.2	98.46 ±0.06	99.42 ±0.04	99.86 ±0.02	99.93 ±0.02	99.92 ±0.02	99.89 ±0.02	99.87 ±0.02	99.91 ±0.04

Table 8.5   Filter Efficiency versus Pressure Drop for 1.0 μm Kinar Filter

ΔP, cm Hg	Dp = 0.034	0.05	0.07	0.10	0.15	0.20	0.30	0.50	0.70	1.0 μm
					Percent Efficiency[a]					
1	b									
3	94.0 ±1.0	93.9 ±0.5	94.2 ±0.2	95.6 ±0.1	98.79 ±0.04	99.19 ±0.05	99.84 ±0.04	99.88 ±0.04	99.86 ±0.02	99.81 ±0.04
10	90.6 ±0.3	91.0 ±0.2	95.47 ±0.09	97.44 ±0.06	99.48 ±0.03	99.78 ±0.03	99.84 ±0.02	99.76 ±0.04	99.69 ±0.05	99.76 ±0.04
20	93.5 ±0.3	94.2 ±0.2	98.74 ±0.08	99.51 ±0.05	99.90 ±0.03	99.91 ±0.02		c		

[a]The efficiencies of two different membranes were measured at p = 10 cm Hg, Dp = 0.20 μm. The measured efficiencies were virtually the same (99.79, 99.76%), although v_0 differed by 40%. (Table 8.1)

[b]Flow rate of sample aerosol too low to permit accurate penetration measurement at this pressure drop.

[c]20 cm Hg obtainable only with one of the three membranes tested.

During the course of the experiments, it was found that there could be considerable variation in the flow resistance for different samples of a given filter media. For instance, in Table 8.1 the minimum and maximum face velocities for three Kinar filters tested are shown. The face velocity is seen to vary by as much as a factor of 1.5 under the same pressure drop conditions. However, the corresponding variation in filter efficiency (Table 8.5) was found to be minimal when the pressure drop across the filter samples was kept the same, but was significant when the same inlet face velocity was maintained. This was thought to be due to the fact that the variation in flow resistance is caused primarily by variations in the pore densities rather than by variations in the filter pore size. Thus, under the same pressure drop conditions, the velocity of the fluid through each pore, and hence the filter efficiency, would be the same. For this reason, the filter pressure drop rather than the inlet face velocity was chosen as the parameter to be varied in the experiments. This results in consistent filter efficiency values in spite of the variation in the individual filter samples.

Some typical experimental results are shown in graphical form in Figures 8.4 and 8.5. When the 0.4-μm Nuclepore filter operates under moderate pressure drops of 3 and 10 cm Hg, the efficiency-versus-particle-size curve does conform to the qualitative prediction of the classical theory: the efficiency is high for both large and small particles, with a minimum at some intermediate particle size. For the specific examples shown, this minimum occurs at particle diameters of 0.085 μm and 0.07 μm. However, as the pressure drop across the filter is increased to 30 cm Hg, the most penetrating (least efficient collection) particle size

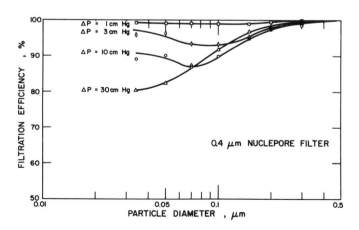

Figure 8.4   Efficiency of 0.4-μm Nuclepore filter.

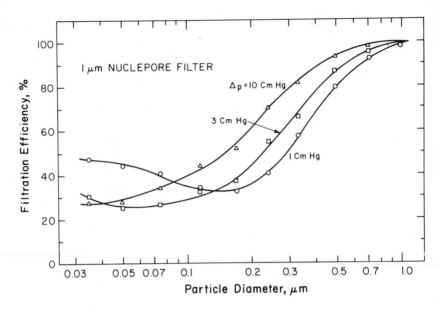

Figure 8.5   Efficiency of 1.0-μm Nuclepore filter.

decreases to below 0.034 μm. The observed filter efficiency in the 0.034 to 1 μm size range then becomes a monotonically increasing function of particle size. Under such conditions, the principal particle collection mechanisms are inertial impaction and interception, with diffusion playing a minor, or perhaps insignificant, role.

The principal experimental results are shown in summary form in Tables 8.6 and 8.7. In Table 8.6 we show the inlet face velocity of the filter at pressure drops of 1 and 10 cm Hg. The inlet face velocity at a given pressure drop provides a measure of the "permeability" of the filter medium. Among the filters tested, the highest permeability is provided by the 5-μm Nuclepore filter, while the lowest permeability is provided by the 0.5-μm Fluoropore filter. More complete filter pressure drop characteristics for some of the filters tested are shown in Figure 8.6.

In Table 8.7 the experimental filter efficiencies are shown for filter pressure drops of 1 and 10 cm Hg. At each pressure drop we show both the minimum filter efficiency in the 0.034-1-μm diameter range and the total filter efficiency for collecting atmospheric fine particles. The latter efficiency was obtained by combining the filter efficiency curves for monodisperse aerosols with the typical size spectrum of atmospheric fine particles shown in Figure 8.7. According to Whitby,[8] the mass distribution

Table 8.6   Flow Resistance of Air Sampling Filter Media

	Face Velocity (cm/sec) at	
	$\Delta P = 1$ cm Hg	$\Delta P = 10$ cm Hg
8 μ Nuclepore	24	210
5 μ Nuclepore	45	450 (extrapolated)
3 μ Nuclepore	20	190
1 μ Nuclepore	22	200
0.6 μ Nuclepore	4.0	40
0.4 μ Nuclepore	4.4	43.3
10 μ Mitex	9	74
5 μ Mitex	4.2	30
1 μ Fluoropore	13	100
0.5 μ Fluoropore	3.0	30
1.2 μ Cellulose (Nuclepore)	–	77.5
1.0 μ Millipore "Kinar"	11.7-16.4	108-154
Pallflex Tissue Quartz (2500 QAD)	21	205.8

Table 8.7   Efficiencies of Air Sampling Filter Media

	$\Delta P = 1$ cm Hg		$\Delta P = 10$ cm Hg	
	η_{min}[a]	η_t[b]	η_{min}[a]	η_t[b]
8 μ Nuclepore	0	9.84	5	51.70
5 μ Nuclepore	0	15.59	–	–
3 μ Nuclepore	8	31.58	10	61.51
1 μ Nuclepore	31	62.34	27	78.78
0.6 μ Nuclepore	81	92.29	47	86.04
0.4 μ Nuclepore	98.9	99.73	87.1	98.15
10 μ Mitex	61	85.27	67	95.92
5 μ Mitex	81	95.10	79	98.59
1 μ Fluoropore	>99.99	>99.99	>99.99	>99.99
0.5 μ Fluoropore	>99.99	>99.99	>99.99	>99.99
1.2 μ Cellulose (Nuclepore)	–	–	>99.6	99.94
1.0 μ Millipore "Kinar"	>99.99	>99.99	90.6	99.54
Pallflex Tissue Quartz	94	99.92	98.1	99.85

[a]η_{min} = minimum efficiency (%) in the 0.034-1.0 μm diameter range

[b]η_t = total efficiency (%) for fine particles (mmd = 0.34 μm and σ_g = 2.04)

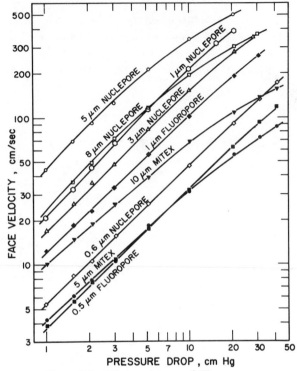

Figure 8.6   Flow resistance of filters.

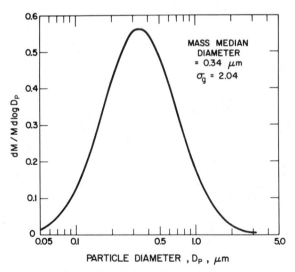

Figure 8.7   Size distribution of atmospheric fine particles used in calculating total efficiency of filters.

of atmospheric aerosols is usually bimodal with a coarse particle mode located at a diameter of 8.75 μm (σ_g = 2.33) and a fine particle mode at 0.34 μm (σ_g = 2.04). By integrating the filter efficiency over the size spectrum of atmospheric fine particles shown, the overall efficiency of the filter for collecting atmospheric fine particles was determined.

ACKNOWLEDGMENT

This research is supported by a grant, #R801301, from the Environmental Protection Agency. The Agency's support is gratefully acknowledged.

REFERENCES

1. Knudson, H. W. and L. White. "Development of Smoke Penetration Meters," NRL Report P-2642 (Washington, D.C.: U.S. Naval Research Laboratory, 1945).
2. Lockhart, L. B., R. L. Patterson and W. L. Anderson. "Characteristics of Air Filter Media Used for Monitoring Airborne Radioactivity," NRL Report 6054 (Washington, D.C.: U.S. Naval Research Laboratory, 1964).
3. Spurny, K. R. and J. P. Lodge. "Collection Efficiency Tables for Membrane Filters Used in the Sampling and Analysis of Aerosols and Hydrosols," Vol. I, II and III, Report No. NCAR-TN/STR-77 (Boulder, Colorado: National Center for Atmopsheric Research, 1972).
4. Liu, B. Y. H. and K. W. Lee. "Efficiency of Membrane and Nuclepore Filters for Submicrometer Aerosols," Environ. Sci. Technol. 10:345 (1976).
5. Liu, B. Y. H. and K. W. Lee. "An Aerosol Generator of High Stability," Am. Ind. Hyg. Assoc. J. 36:861 (1975).
6. Dzubay, T. G. and R. K. Stevens. "Ambient Air Analysis with Dichotomous Sampler and X-Ray Fluorescence Spectrometer," Environ. Sci. Technol. 9:663 (1975).
7. Loo, B. W., J. M. Jaklevic and F. S. Goulding. "Dichotomous Virtual Impactors for Large Scale Monitoring of Airborne Particulate Matter," in Fine Particles, B. Y. H. Liu, Ed. (New York: Academic Press, 1976), pp. 312-350.
8. Whitby, K. T. "Modeling of Atmopsheric Aerosol Particle Size Distributions," Progress Report, EPA Grant R800971, Particle Technology Laboratory Publication No. 253, Mech. Eng. Dept. (Minneapolis: University of Minnesota, 1975).

9

COLLECTION SURFACES OF CASCADE IMPACTORS

J. J. Wesolowski, W. John and W. Devor

Air and Industrial Hygiene Laboratory
California Department of Health
Berkeley, California

T. A. Cahill, P. J. Feeney, G. Wolfe and R. Flocchini

Crocker Nuclear Laboratory and the
Department of Physics
University of California
Davis, California

INTRODUCTION

At present, most methods used to measure particle size distributions of specific chemical species in ambient air require a two-step process. First, an instrument is used to size-segregate the particles onto suitable collection surfaces. These surfaces are then sent to the laboratory where appropriate analyses are carried out to quantitate the chemical species. The total error associated with the measurements involves both the collection and analytical errors, the former often being the larger.[1-5] It should be recognized that much data were obtained in the past by size-segregating instruments using surfaces whose collection properties were not known. In many cases these data are in serious error.

An important collection error is that associated with "bounce off" of particles from collection surfaces. A particle striking a surface can stick to the surface, bounce off, break into fragments, or dislodge other particles previously collected. The word *reentrainment* is used to refer to this last process as well as to the loss of previously collected particles caused by the motion of the air alone. It is difficult to separate these phenomena

121

experimentally, and therefore the term *bounce off* is often used to refer to the general problem. A particle reentering the aerosol stream will either be lost to the walls of the device or be collected by a subsequent impaction stage or the after-filter. (The latter phenomenon is referred to as cross-over.) The word *sticky* is used to refer to surfaces that have a good collection efficiency. The ability of surfaces to absorb the kinetic energy of the particles may be as important for good collection efficiency as their ability to wet the particles.

The type of collection surface to be used will depend upon the subsequent chemical analysis. A surface covered with grease (*e.g.,* silicone high vacuum grease) provides good collection efficiency and is suitable for some purposes, but it makes gravimetric mass determinations difficult and increases blank values for some chemicals of interest. Furthermore with analytical techniques such as electron spectroscopy for chemical analysis (ESCA), which analyzes only the upper 30-50 angstroms of surfaces, a medium providing any shielding of the particles must be avoided. For techniques based on the X-ray fluorescence principle, problems associated with self-absorption of the X-rays must be considered, particularly for the light elements. Such absorption increases for particles imbedded within a collection surface such as a grease.

The purpose of this chapter is to present both laboratory and field measurements employing various surfaces, and to make specific recommendations concerning the choice of collection surfaces.

THE MULTI-DAY CASCADE IMPACTOR

The most commonly used instrument for size-selective sampling is the cascade impactor because the particles are aerodynamically separated and the samples obtained can be taken to the laboratory for chemical analysis. The instrument used in the experiment discussed below is the Multi-day Cascade Impactor,* which has been suggested as a possible monitoring device.[6] It is a modification of the rotating drum Lundgren impactor, which is frequently used in research experiments to obtain the particle size distributions of specific chemicals as a function of time.[1]

A schematic of the Multi-day is shown in Figure 9.1. The intake manifold, which has a cut-off of about 20 μm, is followed by two impaction stages and an after-filter. Each stage consists of a rotating cylinder that can be covered with a removable collection surface. The incoming air

---

*Manufactured by Sierra Instruments, Carmel Valley, California. Mention of commercial products, their source or their use in connection with material reported herein is not to be construed as either an actual or implied endorsement of such products.

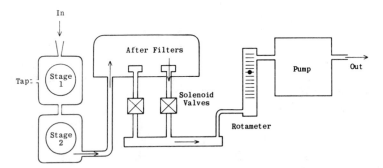

Figure 9.1   Flow schematic of the Multi-day Impactor.

passes through a rectangular slit, and the resulting air jet is incident on the surface of the drum. Large particles with sufficient inertia will leave the streamline and impact on the drum. Smaller particles will follow the streamline to the second stage, which has a smaller slit with a higher jet velocity, and hence will impact on the second drum. Those particles too small to be collected by either drum are collected by the after-filter.

In the monitoring mode, the cylinders are rotated once every seven days. Thus, the collection surfaces can be cut into seven sections each with a 24-hour deposit of size-segregated particles. The after-filter assembly consists of a set of after-filters in separate holders. One filter is activated every 24 hours. Design criteria for a flow rate of 28.3 liters/min lead to 50% cut-off diameters (for unit density spheres) of 3.6 and 0.65 μm for stages 1 and 2 respectively.

EXPERIMENTAL PROCEDURES AND RESULTS

Laboratory Measurements

The laboratory experiments were carried out at the Air and Industrial Hygiene Laboratory (AIHL) Aerosol Physics Laboratory in Berkeley, California. Four types of collection surfaces were studied: bare Mylar 0.0025 cm thick, Mylar covered with a layer of paraffin wax approximately 50 μg/cm² thick, and Mylar covered with two grease coatings: Dow Corning silicone high vacuum grease and Apiezon (Type M) grease. The paraffin coating was formed by dipping 0.0025-cm thick Mylar foils into a 2.5% solution of paraffin wax in toluene. The silicone grease was applied in a relatively thick layer using a paper tissue (Kimwipe). The excess was then wiped off with a clean tissue, leaving a smooth, thin layer tacky to the touch with a thickness greater than several particle diameters. Apiezon grease was applied by two different methods. The first was identical to the above method for silicone, and the second consisted of dipping the drum in a saturated solution in toluene for five seconds.

Monodisperse methylene blue aerosols were generated by a Berglund-Liu vibrating orifice generator.[7]   In this generator a solution of methylene blue in isopropyl alcohol and water is forced through a 20-μm orifice that is vibrated ultrasonically.   This causes the jet to break into uniformly sized droplets that are dried in dilution air to produce monodisperse, spherical particles.   Electrical charges on the particles are neutralized by a Kr-85 radioactive source.   The experimental arrangement, including an optical counter used to monitor the aerosol size distribution and intensity, is shown in Figure 9.2.

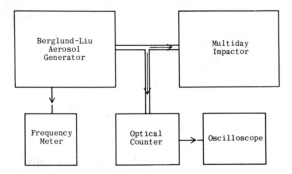

Figure 9.2 Experimental arrangement for measurements with methylene blue aerosol.

Experiments were carried out with 6 μm and 1.5 μm particles.   The former are expected to be collected primarily by stage 1 of the impactor and the latter by stage 2.   Methylene blue particles are spherical, smooth, elastic and hence expected to be "bouncy."   Figures 9.3 (a, b, c and d) qualitatively demonstrates bounce off as a function of collection surface for the 6 μm particles on stage 1.   Each of the surfaces received the same exposure to methylene blue aerosol.   Figure 9.3a shows the results for Dow Corning silicone vacuum grease.   There are indications of some bounce off, although some of it may be due to having too heavy a deposit, so that the particles contact the previously deposited methylene blue rather than the silicone grease.   Figure 9.3b illustrates results for use of Apiezon vacuum grease.   It is rather good in collecting the particles, but not quite as good as silicone.   Figure 9.3c demonstrates the large loss from the paraffinated Mylar, while 9.3d shows practically no deposit on the bare Mylar.

a

c

b

d

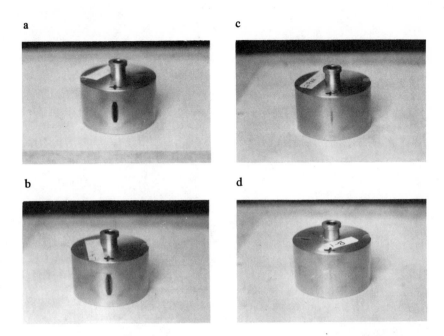

Figure 9.3   Deposits of 6-µm methylene blue particles on various surfaces. All received equal exposures. (a) silicone vacuum grease; (b) Apiezon vacuum grease; (c) paraffin wax; (d) bare Mylar.

Figure 9.4 (a, b, c and d) shows the same sequence of collection surfaces for 1.5-µm particles incident on the second stage where they should essentially all impact. Figure 9.4a shows the heavy deposit obtained with silicone grease. Note that the slit is narrower on this stage than on stage 1. The asymmetric pattern of deposited particles is probably due to the off-center location of the vacuum outlet. As mentioned earlier the drums are rotated in field operation to obtain a time history of the deposit. Therefore bounce off can cause errors in the time history by depositing particles on the wrong area of the drum. The relative results for Apiezon, paraffin and bare Mylar shown in Figures 9.4b, c and d are similar to those presented for the 6-µm particles on stage 1.

Figure 9.5 shows the measured collection efficiencies for the surfaces obtained by dissolving the methylene blue in an alcohol-water solution and quantitating on a spectrophotometer. The total was obtained by summing the stage in question and any later stages, after-filter and walls. The amount on the walls was obtained by washing the inside of the impactor. It should be emphasized that most of the missing material was

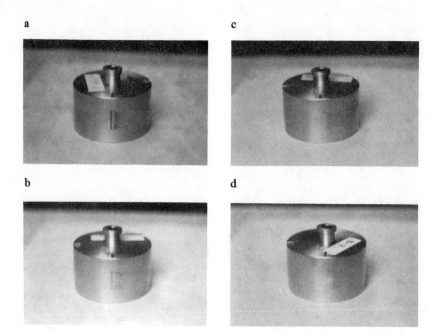

Figure 9.4   Deposits of 1.5-μm methylene blue particles on various surfaces. All received equal exposures. (a) silicone vacuum grease; (b) Apiezon vacuum grease (the double slit image was caused by a movement of the drum); (c) paraffin wax; (d) bare Mylar.

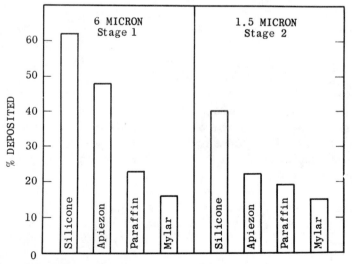

Figure 9.5   Sticking efficiency of various substrates for methylene blue particles. The less than 100% values for silicone and Apiezon are attributed to heavy deposits of methylene blue.

found on the walls. The wall losses for impactors quoted in the literature are usually based on tests with dioctyl phthalate (DOP), a sticky oil aerosol. Since there is less bounce off with DOP, wall losses are also expected to be lower than those obtained with methylene blue, i.e., the wall loss is a function of the aerosol used. Although the sticking efficiencies for silicone and Apiezon are considerably larger than for paraffin and Mylar, the values for silicone and Apiezon in Figure 9.5 are appreciably less than 100%. This is attributed to heavy deposits in this measurement resulting in coating of the greases with a layer of methylene blue.

A. K. Rao has recently evaluated a number of surfaces using polystyrene latex particles of approximately 1 μm diameter.[8] For silicone oil and petroleum jelly, he found nearly 100% sticking efficiency, whereas uncoated glass plates gave efficiencies only of the order of 20%. He also evaluated filters used as impaction surfaces. His results show that filters help to retain large particles although they are not as good as greases or oils. However, the filters also stop some of the small particles, thus spoiling the cut-off characteristics of the stage. This can be attributed to the fact that the air jet penetrates partially into the filter where some of the small particles are trapped.

Recent work by Dzubay and Stevens using ambient air aerosol illustrates another consequence of using substrates of poor collection efficiency.[2] With two impactors in parallel, one using greased surfaces and one using bare aluminum, they found that the mass median diameter of an aerosol shifted from 8 μm for the greased surface to 3.5 μm for the bare aluminum.

Field Measurements

Ambient air sampling was carried out by the staff of the University of California, Davis, in the summer and fall of 1975 during periods of low relative humidity (< 50%) and high daily maximum temperatures (up to 40°C). Such conditions are expected to produce dry, soil-derived aerosols that should provide an extreme test of the collection efficiency of cascade impactor collection surfaces. Sampling was carried out at two locations, the Air Resources Board (ARB) monitoring station at 1025 P Street in Sacramento, and the roof of the Physics-Geology building at the University of California, Davis. Particulate matter was collected for a total of 72 hours. Alpha-induced X-ray emission analysis using the U.C. Davis isochronous cyclotron was used to determine the concentration of the elements on the collection surfaces and filters.

Tables 9.1 and 9.2 present data on the relative collection efficiency for different surfaces, the wall losses in the impactor, and the stage cross-over. The data were obtained by running a parallel array of Multi-day impactor systems, each with a different collection surface, and a total filter that had an intake manifold similar to that of the impactors. Prior to these runs the equivalency of the Multi-day units under field conditions was determined by running a parallel array with identical collection surfaces at the Sacramento location. Five units were spaced in a grid with two-meter intervals near the middle of the roof area and as far from walls and outlet vents as possible.

Flow rates were measured both by hot wire airflow indicators and by the pressure drop across the first stage as measured with a magnehelic gauge. Results of the two flow measurements agreed to within ± 5% for all measurements both before and after the 24-hour samples were collected. Elemental analysis was carried out for the major elements on collection surfaces from all the units. The agreement for the major elements was better than ± 10% for three of the units, which were then used in future runs for the paraffin, Apiezon and silicone coatings. The other two units showed variations as large as ± 20%. One of these was used on later runs for the bare Mylar coatings.

Table 9.1 lists the data for the relative collection efficiencies for the various surfaces. The data are clustered into four categories. Elements with statistically similar size distributions and highly correlated elemental concentrations are gathered into subsets. The soil subset (Al, Si and Fe) consists of highly correlated elements (> 0.8) from large particles. The automotive subset (Br, Pb) possesses correlations greater than 0.95 and has size distributions dominated by small particles. The sulfur size distribution was dominated by small particles but also had a significant component in the intermediate size range.

Table 9.1 Relative Collection Efficiency (%) of Different Collection Surfaces for Particles in Ambient Air[a]

Elemental Subset	Bare Mylar	Paraffin	Silicone	Apiezon
Soil[b] (stage 1)	5	44	93	100
Na, Cl (stage 1 and 2)	38	89	89	100
S (stage 1 and 2)	<9	35	70	100
Br, Pb (stage 1 and 2)	9	51	92	100

[a]See text for explanation of percentage calculation.
[b]Soil-derived aerosols include elements that were strongly correlated (> 0.8) with silicon, namely Al, Si and Fe.

Relative collection efficiency was calculated by taking the highest measured value for each element and each stage for the four collection surfaces, setting it equal to 100%, and forming percentages for the remaining three. These values were then averaged over elements in the given categories. For Al, Si and Fe only stage 1 was used, since most of the total mass for these elements was collected on stage 1.

The total analytical errors associated with these ambient air measurements varied from a few percent to as high as 50% depending upon the element and its concentration. Hence great significance should not be placed in small differences in the numbers in Table 9.1.

The data shown in Table 9.1 for the soil set demonstrate that Apiezon and silicone greases are far more efficient collectors of soil-derived aerosol than paraffin, while Mylar is an extremely poor collection surface. This is consistent with the methylene blue particle data discussed earlier. Since sodium and chlorine did not correlate well with the soil set, they are from other sources, either anthropogenic or natural, such as sea salt aerosol transported into the valley. Since sea salt is hygroscopic, it is not surprising to find that the paraffin and bare Mylar have higher collection efficiencies for sodium and chlorine than for the soil set. This is consistent with work done in the aerosol characterization experiment (ACHEX).[3] In that experiment, two Lundgren-type impactors, one with bare Mylar collection surfaces and one with sticky surfaces, were run in parallel. The sampling, carried out in a near-coastal location (Berkeley, California), demonstrated the sodium distributions to be identical, within experimental errors, for the two surfaces.

Table 9.2 lists wall losses and cross-over of soil-derived silicon particles. The first line lists the wall losses, *i.e.,* the percentage of the particles entering the impactor that were not collected by either the stages or the after-filter. These numbers were obtained by comparing the amount of silicon measured on the total filter (which was run in parallel with the Multi-day) with the sum of the silicon on the stages and after-filter. The

Table 9.2   Wall Losses and Cross-Over of Soil-Derived Silicon Particles (%)

	Bare Mylar	Paraffin	Silicone	Apiezon
Si wall losses in impactor	65	34	$< 5^a$	< 1
Si on after-filter as % of Si on sum of stages 1, 2 and after-filter	81	18	< 2	< 2

[a]Larger uncertainty due to blank subtraction for silicone-coated Mylar.

wall losses for the paraffin and Mylar are large and are similar to those determined in the methylene blue experiment discussed earlier. They also are very similar to the data reported in the ACHEX experiments for wall losses of aluminum particles collected on bare plastic surfaces. In that experiment bare Teflon and Mylar surfaces were shown to have wall losses of 50% and 70%, respectively.[4]

Line two in Table 9.2 gives the amount of silicon captured on the after-filter as a percentage of the sum of the silicon captured on stage 1, stage 2 and the after-filter. This gives a measure of the cross-over. The silicone and Apiezon grease data indicate that practically none of the silicon should be collected on the after-filter. The fact that in the bare Mylar case 81% of the silicon was on the after-filter compared to only 18% in the case of paraffin confirms that the bare Mylar has greater cross-over than the paraffin.

CONCLUSIONS

The following conclusions and recommendations can be made as a result of the above work.

1. Based on both laboratory and field work, silicone and Apiezon greases are found to have much higher collection efficiencies than either paraffin wax or bare Mylar. It is therefore recommended that greased collection surfaces be used, although it is recognized that total mass data cannot be obtained without serious difficulty nor can certain elements be determined due to high blank values.

2. For many elements, especially those that are soil derived, the wall loss and cross-over effects are sufficiently large from paraffin and bare Mylar collection surfaces that data collected using such surfaces are seriously in error.

3. Although the data presented were obtained with a particular type of cascade impactor, we expect that collection surface problems of the type found here are inherent in most impaction systems. Thus no impactor should be used for research or monitoring purposes unless experiments similar to those discussed here are carried out to determine the validity of the data.

4. It is recommended that more work be done to develop collection surfaces that have high collection efficiency, low chemical blank values, and that can be used for mass determinations. We also recommend that other types of size-segregating instruments be evaluated, such as the virtual impactor and the cyclone.

5. The fact that the laboratory and field evaluations of substrates led to similar conclusions concerning their collection efficiencies implies that laboratory measurements are very useful for selecting collection surfaces before final testing in the field.

ACKNOWLEDGMENTS

This work was funded by the California Air Resources Board. We appreciate the interest and support of Dr. Jack Suder.

REFERENCES

1. Lundgren, D. A. "An Aerosol Sampler for Determination of Particle Concentration as a Function of Size and Time," *J. Air Pollution Control Assoc.* 17:225 (1967).
2. Dzubay, T. G., L. E. Hines and R. K. Stevens. "Particle Bounce Errors in Cascade Impactors," *Atmos. Environ.* 10:229 (1976).
3. Wesolowski, J. J., A. E. Alcocer and B. R. Appel. "The Validation of the Lundgren Impactor," California State Department of Health, AIHL Report No. 138-B (September 1975). Accepted for publication in the ACHEX/Hutchinson Memorial Volume, Ann Arbor Press.
4. Wesolowski, J. J. *Second Joint Conference on the Sensing of Environmental Pollutants* (Pittsburg, Pennsylvania: Instrument Society of America, 1973), p. 191.
5. Cahill, T. A. and P. J. Feeney. "Contribution of Freeway Traffic to Airborne Particulate Matter," Final Report UCD-CNL 167 (Sacramento, California: California Air Resources Board, 1973).
6. Flocchini, R. G., T. A. Cahill, D. J. Shadoan, S. F. Lange, R. A. Eldred, P. J. Feeney, G. Wolfe, D. C. Simmeroth and J. K. Suder. "Monitoring California's Aerosols by Size and Elemental Composition," *Environ. Sci. Technol.* 10:76 (1976).
7. Berglund, R. N. and B. Y. H. Liu. "Generation of Monodisperse Aerosol Standards," *Environ. Sci. Technol.* 7:147 (1973).
8. Rao, A. K. "An Experimental Study of Inertial Impactors," Ph.D. Thesis. (Minneapolis, Minnesota: Particle Technology Laboratory, University of Minnesota, 1975).

SECTION III

SAMPLE PREPARATION FOR WATER ANALYSIS

10

SAMPLE PREPARATION FOR
MULTIELEMENTAL ANALYSIS OF WATER

F. A. Rickey, K. Mueller, P. C. Simms and B. D. Michael
Department of Physics
Purdue University
West Lafayette, Indiana

INTRODUCTION

Through improved techniques for performing multielemental analysis, such as X-ray fluorescence (XRF) and proton-induced X-ray emission (PIXE), we learn new information every day about our environment. When these techniques are applied to the analysis of the liquid portions of our environment, the problem of sample preparation is created. Before water samples can be analyzed by XRF or PIXE the water must be eliminated; that is a thin dry film consisting of the trace impurities deposited on a suitable substrate must be prepared. The particular technique used in our laboratory to analyze water samples is PIXE, and the substrate must be a thin film suitable as a backing for a target exposed to a proton beam. In order to reach sensitivity levels of 0.1-1.0 ppb (current standards for water quality studies) approximately 20-40 ml of solute must be reduced to dryness without significant addition or removal of trace elements.

There are several problem areas in this process. First, the effects of kinetics must be minimized. For example, in a simple boiling process large amounts of material can be lost when drops of solution are lost from the surface of the sample. The same sort of effect occurs in a standard freeze-drying process if the sample is in the form of a large chunk of ice. Second, a closed system must be used to minimize contamination from atmospheric dust particles entering an open system. In our laboratory, with all of its galvanized ducts and brass fittings, a dust particle is likely to contain large amounts of iron, zinc, tin and copper. We have measured 100 ppb contaminations that result from vessels being left open. Third, one must quantitatively transfer the solids from the concentration unit to a suitable substrate. Imagine the challenge of transferring a milligram of material to a backing without leaving a microgram or so behind. If this microgram contains most of the cadmium in the sample, then the analysis is doomed. Finally, the trace elements in the sample may exist in many varieties of chemical forms. Thus one desires a preparation technique independent of the chemistry of the sample.

While struggling to find successful solutions to these problems we came up with a novel approach. Consider constructing a container whose bottom surface is permeable to water vapor but not water, and exposing the bottom surface to a suitable vacuum. Any water in the container cannot pass through the membrane, and only water vapor is pumped through the membrane. In this conceptual system the water level drops as vapor is pumped away, and all of the dissolved solids are left behind. Since the only place at which the water changes state is at the membrane, hopefully all the solids end up on or in the membrane. Finally, the membrane itself may serve as a suitable backing for a target to be used in PIXE analysis. This concept, which we call *vapor filtration*, has many attractive features. The water is not agitated, so there is no danger of water droplets spattering with subsequent loss of material. The system can be closed at the top so that airborne contaminants cannot enter. Because there is no final transfer of solids involved, the process can be quantitative. Because no chemical processing is required, the process is not affected by the chemical composition of the dissolved solids.

EXPERIMENTAL METHODOLOGY

The first question is obviously whether a suitable membrane exists. Happily there are several membranes that are permeable to water vapor.[1] They are polymers consisting of carbon, oxygen and hydrogen, and thus are, in principle, good candidates for target backings. The membranes are sturdy when dry. However, when they become saturated with water vapor, their mechanical strength drops markedly, and the membrane must be supported with some porous material like fritted glass.

Once a membrane has been selected it must be incorporated into a suitable holder. The details of this are not terribly critical; all that is required is (1) a good liquid seal between the membrane and the water container, (2) a good vacuum seal between the membrane and the vacuum system, and (3) a porous support for the membrane. If one is interested in a limited number of samples, then commercially available Millipore filter

holders can be used.   These holders consist of two glass cylinders with a
flat ground-glass joint.   The bottom unit has a fritted glass plug in the
plane of the joint.   The membrane may simply be clamped between the
two pieces.   There are drawbacks to this system, however.   Glass is very
hard to clean, it breaks easily, and the units are expensive.

   We have constructed our own units, and the essentials of the assembly
are shown in Figure 10.1.   The top piece is a plastic cylinder glued into

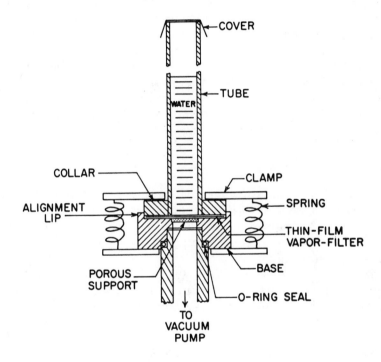

Figure 10.1   Diagram of vapor filtrations unit.

a clamping disk.   The base is cut from plastic rod, and machined with an
axial hole for pumping.   The base has a recess for insertion of a polyethylene frit, an O ring seal to a vacuum port, and a lip on the outside
top for ready alignment of the circular membrane.   The top and base are
held together by a spring-loaded clamp.   Medicine cups are used for the
top cover.   Only the top cylinder is in contact with the water for any
length of time, and it is easily cleaned.   Since we are interested in the
simultaneous preparation of many samples, a vacuum manifold has been
constructed that has 120 pumping ports.

The vacuum system has special requirements. Care must be taken to keep the pressure below 4 Torr, which is the vapor pressure of ice at 0°C. The membrane becomes cold, and water vapor. passing through the membrane could recondense on the back side of the membrane and instantaneously freeze. Ice is not permeable to water vapor. If the pressure remains below 4 Torr, water cannot exist on the back of the membrane. While this is not a difficult task, under normal circumstances a standard mechanical pump will not pump water vapor. One solution is to cryogenically pump the water vapor with a cold trap. This has the disadvantages of (1) letting water vapor into the mechanical pump at a slow rate, which will eventually stop the pump, and (2) defrosting the trap periodically, which means the system must be shut down. The pumping system we use consists of a large mechanical pump whose outlet pressure is maintained at 75-100 Torr by a water-ring pump. The water vapor is continuously pumped from the warm pump oil by the ring pump, which is an efficient water vapor pump in the 75-100 Torr range.

EXPERIMENTAL RESULTS

The vapor filtration rate depends on the thickness and type of membrane used and the total quantity of dissolved solids. Typically one can remove about 0.7 ml of water per hour. Thus, sample preparation must start several days before analysis. The second general observation is that the deposit on the membrane is not uniform. This nonuniformity varies with the solids content of the sample, and within the deposit the distribution of elements is not homogeneous. For PIXE analysis, this problem can be solved by bombarding with a uniform beam that entirely covers the target spot. We use a beam that is uniform to within 5%. This remaining nonuniformity is a source of uncertainty in our analysis; we believe the total uncertainty is less than 10%.

In order to perform quantitative tests of the target preparation technique, one must have a reliable calibration of the analytical technique. This problem is discussed in Chapters 12 through 15 of this book. Our initial calibration was based on the X-ray yields in $(count/\mu g)/(\mu Coul/cm^2)$ measured for evaporated metal-foil calibrators purchased from Micro-Matter Co.[2]   One must also have water samples available whose elemental composition is known. This is not a trivial requirement! We approached this problem by purchasing an ultrapure water system from Culligan Corporation. Next we purchased a series of atomic absorption (AA) standards that are certified to contain 1000 ppm of various single elements in aqueous solution. Our first test samples were single-element samples prepared by adding appropriate quantities of the AA standards to clean water to produce concentrations varying from 5 ppb to 2 ppm. These samples were passed through the vapor-filtration system and analyzed using the PIXE technique.

Figure 10.2 shows a comparison of the results from liquid samples and evaporated metal-foil calibrators as a function of atomic number Z.

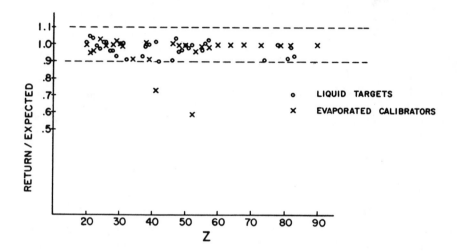

Figure 10.2 Comparison of evaporated calibrators and liquid single-element targets, prepared by vapor filtration technique. The measured yields are divided by the best-fit expected yields.

Notice that in no case does the measured return from a liquid sample deviate by more than 10% from the average yield obtained by fitting a smooth curve to the data from all calibrators. Two of the evaporated metal-foil calibrators, Nb and Te, were obviously faulty. Figure 10.3 shows a yield curve for K_α X-rays as a function of Z, using both sets of data. For most elements, the liquid and evaporated calibrators were in such good agreement that they could not be plotted separately. As shown in Figure 10.3, it is easy to distinguish between the calibrators that are slightly low and the majority that are on the smooth curve. These excellent results for single-element samples in water were equally good over a range of concentrations from 5 ppb to 2 ppm.

Single-element samples are a far cry from real water, and we were worried about possible interferences between elements in a multielement sample. For example, two elements might coprecipitate early in the process so that losses to the cylinder walls might be significant, or a combination of elements might change the mechanism of the membrane.

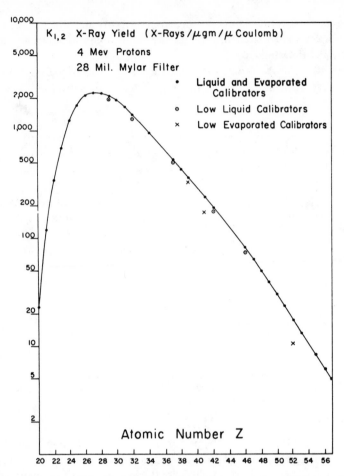

Figure 10.3   Use of liquid single-element targets as an independent test of the PIXE calibration curve.

While preparing single-element samples is straightforward, preparing multi-element samples is not. Mixing the wrong two AA standards can result in very unpredictable concentrations, due to chemical reactions and pre-cipitations. We separated 40 AA standards into four groups, and prepared multielement targets that seemed to be stable. Figure 10.4 shows the results from one of these groups, which are typical of results from the other three. For example, copper gives a slightly smaller return than the others, but it is within 10%. Indeed, considering the problems involved in making stable samples it seems that the vapor-filtration system has performed admirably.

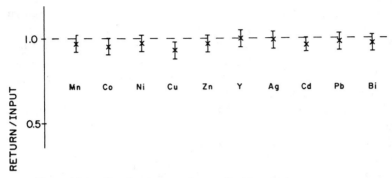

Figure 10.4   Fractional return from a liquid multielement target.

These tests indicate that losses to the cylinder walls are negligible. The preceding tests were somewhat unrealistic, however, in that relatively small concentrations of dissolved solids were present. We investigated the possibility that at high concentrations sodium or calcium might plate on the cylinder walls, and that other elements might coprecipitate with the sodium or calcium. Accordingly we prepared multielement samples with varying amounts of calcium and sodium added. For sodium or calcium concentrations varying from 0 to 200 ppm we could see no effect within 10%. Figure 10.5 shows two examples, involving zinc and silver. These examples were selected because of the low absorption of zinc and silver K X-rays in a calcium matrix.

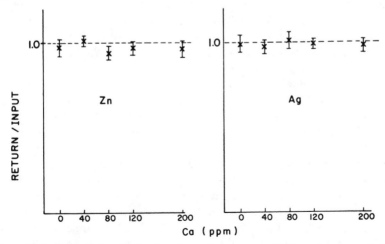

Figure 10.5   Test of the effects of various calcium concentrations on return of zinc and silver.

The calcium tests were also run for titanium, chromium, and manganese, where absorption effects might be severe. If these elements were uniformly distributed with the calcium so that on the average the X-rays had to pass through half of the calcium, there should be decreases in X-ray yields of 35%, 22% and 18% for titanium, chromium and manganese, respectively, when calcium is present at concentrations of 200 ppm. Surprisingly no effects greater than 10% were observed. Limited tests with the calcium concentration increased to 800 ppm were made, with the same results. Thus it seems that these elements are deposited on top of the calcium, which was an unexpected bonus. We have measured significant absorption of barium L X-rays, but the correct yields can be extracted from the K X-rays.

One of the goals set forth was to develop a process in which samples were not contaminated. This question of contamination was investigated by a series of tests in which the samples were placed in ultrapure water (acidified to a pH of 2.3 with HNO_3 to simulate actual samples). The water itself was quite clean, containing less than 1 ppb of even the most abundant impurities. As we had hoped, samples could be prepared without introducing contaminants from the air. We did find, however, that some elements are harder to clean from the tubes than others. One of the tubes we tested, for example, had been used to prepare a 2-ppm single element bromine sample. After the tube had been washed with a 10% HNO_3 solution we still picked up a substantial bromine residue. (In our worst case, we observed about 20 ppb.) Cleaning this tube in an ultrasonic cleaner removed all traces. Tubes used for high level iron and calcium tests, on the other hand, were very clean. Generally speaking, a standard wash with 10% HNO_3 is sufficient to reduce residues to acceptable levels. If, however, large amounts of a few elements have been introduced into the tubes, extra cleaning is required. We, in fact, have set aside tubes to be used only for high level tests. In addition, all tubes are cleaned periodically in the ultrasonic bath.

Every test we have performed indicates that the vapor-filtration scheme for preparing water samples satisfies the requirements set forth. It is relatively simple, it is clean, and it is quantitative. Purdue University has applied for a patent on the technique and device used. Since the system was developed we entered into a contract with the EPA[3] to analyze a large number of water samples. It is one thing to prepare a few samples on a pilot system where the routine procedures are handled by trained scientists. It is altogether different to prepare large numbers of widely different samples, particularly if a technician must handle all the procedures. Accordingly, we have had to standardize all techniques and provide for built-in checks to detect mistakes. As an example of how the system can be used in a large scale project, the last section summarizes our operating procedure.

OPERATING PROCEDURE

The water samples are prepared in a clean room that is kept at a positive pressure and whose incoming air is filtered. The cleaning station operates in this room. The tubes are first rinsed for 30 minutes in circulating distilled water, then soaked for 30 minutes in 10% HNO_3, and finally rinsed for 30 minutes in ultrapure water. The components are then stored in closed containers until use. The 120-port vacuum manifold is also in this room. When sample preparation begins, the vapor filtration unit is assembled and the sample poured in (no pipetting). We monitor for system leaks by introducing yttrium as an integral standard before covering the tubes. We routinely prepare samples of pure water in parallel to monitor contamination. Water levels are checked periodically, and the samples removed and stored when dry. In this routine application the success rate has been high, and our experience is that the vapor-filtration system is a practical solution to problems of sample preparation of water for multielement analysis.

REFERENCES

1. Covered in patent application.
2. Micro Matter Co., 177 34th Avenue East, Seattle, Washington 98102.
3. Environmental Protection Agency, Contract No. 68-03-2178.

11

PRECONCENTRATION OF IONS FOR
X-RAY DETERMINATION USING IMMOBILIZED REAGENTS

Donald E. Leyden*

Department of Chemistry
University of Georgia
Athens, Georgia

INTRODUCTION

The potential for the application of X-ray fluorescence to the determination of trace elements in water is similar in practice to that for aerosols. Unfortunately, X-ray fluorescence is not inherently a trace technique by modern standards. We attempt to extend the range of X-ray fluorescence by improvements in the instrumentation, or by some form of sample preconcentration. It is worthwhile to compare briefly the problems and advantages of X-ray fluorescence when applied to aerosol and water samples.

In this book, several practical sampling methods for aerosols have been discussed. Preconcentration of these samples is achieved by filtration of large volumes of air. There is no practical difficulty in sampling one cubic meter or more of air. Therefore large preconcentration factors can be achieved. However, there is serious objection to sampling more than one liter of water, particularly if the water sample must be transported to a laboratory for analysis. On the other hand, preparation of reliable standards for aerosols is a difficult task, whereas preparation of standards for water samples is relatively easy.

The main result of these comparisons is that successful application of X-ray fluorescence methods to determination of dissolved ions in water will depend upon the development of a reliable technique for sampling and preconcentration. An ideal technique should provide *in situ* sampling and

---

*Present address: Department of Chemistry, University of Denver, Denver, Colorado.

preconcentration resulting in a form suitable for X-ray fluorescence analysis. Methods using ion exchange resin-impregnated filter papers have been presented previously.[1,2] Methods based upon the use of ion-exchange resin beads,[3] chelating groups immobilized in polyurethane foam[4] and precipitation[5] have been reported in the literature. This chapter reports an alternative approach using simple chemical reactions to attach chemical functional groups to solid substrates such as silica or porous glass beads.

There are several advantages to this approach. First, the chemistry involved is very simple and many useful reagents are commercially available. Second, the use of chelating groups attached to the glass surface provides for efficient and rapid recovery of metal ions (and in some cases anions) from aqueous solution. After mixing the beads with powdered cellulose, pellets are pressed easily. This method provides a reproducible matrix for the ions with few interelement effects. Standards are prepared by treating known solutions in a way identical to the sample. If the presence of strong chelating groups is suspected in the sample, the method of standard additions may be used to establish concentrations in the samples.

A potential disadvantage of the method is that the pellet prepared is a "thick" sample for the X-rays emitted by some elements and a certain attenuation of the signal results in less than maximum sensitivity. The silicon in the glass causes the matrix to have a rather high mass absorption coefficient for X-rays commonly used for elemental determinations of dissolved ions. However, the general experience in our laboratory is that detection limits of 0.05 μmol of elements such as manganese, iron, cobalt, nickel, copper, zinc, silver and cadmium are obtained.

PREPARATION OF GLASS BEADS

The reagents used are silylation reagents, usually obtained as a trimethoxysilane derivative, available from Dow-Corning or Union Carbide Corporation. For illustration, one such reagent will be discussed in detail because it has found direct application in our laboratory. N-β-Aminoethyl-γ-aminopropyltrimethoxysilane (Dow-Corning Z-6020) contains a 1,2-diamine terminal group that is effective for chelation of many metal ions. In addition, this functional group is easily converted to a dithiocarbamate derivative that is very effective as a chelating group for transition metal ions. A representation of the reactions involved is as follows.

$$\text{Glass Surface}\begin{array}{l}\text{—OH}\\\text{—OH}\\\text{—OH}\\\text{—OH}\end{array} + (CH_3O)_3SiCH_2CH_2CH_2\overset{H}{-N}-CH_2CH_2NH_2 \longrightarrow$$

$$\begin{array}{l}\text{—OH}\\\text{—OH}\\\text{—O—}\\\text{—OH}\end{array}\overset{|}{\underset{|}{\overset{O}{\underset{O}{Si}}}}CH_2CH_2CH_2\overset{H}{N}-CH_2CH_2NH_2 + 3CH_3OH \quad (11.1)$$

$$\begin{array}{l}\text{—OH}\\\text{—OH}\\\text{—O—}\\\text{—OH}\end{array}\overset{|}{\underset{|}{\overset{O}{\underset{O}{Si}}}}CH_2CH_2CH_2\overset{H}{N}-CH_2CH_2NH_2 + CS_2 \quad \xrightarrow{\text{base}}$$

$$\begin{array}{l}\text{—OH}\\\text{—OH}\\\text{—O—}\\\text{—OH}\end{array}\overset{|}{\underset{|}{\overset{O}{\underset{O}{Si}}}}CH_2CH_2CH_2\overset{H}{N}CH_2CH_2N\overset{H}{C}\underset{S^-}{\overset{S}{\diagup}} + H^+ \quad (11.2)$$

Reaction 11.1 is accomplished by stirring for ten minutes 50 g of glass beads (Electro-Nucleonics, CPG-10,200-400 mesh) with a solution prepared from 10 ml Dow-Corning Z-6020, 18 ml 0.1% aqueous acetic acid and sufficient deionized water to bring the total volume to 100 ml. The beads are filtered from the reaction mixture and dried for 12 hours at 80°C. The product is stable unless treated with solutions above pH 10 or below pH 3.

A dithiocarbamate derivative is prepared according to Reaction 11.2 by stirring 30 g of the glass treated as described above with a solution containing 100 ml benzene, 20 ml 2-propanol, 20 ml CS$_2$ and 5 ml of a 10% methanolic solution of tetramethylammonium hydroxide. After 15 minutes, the beads are filtered from the reaction mixture, washed with 2-propanol and air dried. This product is stable for several days. However, since discoloration resulting from oxidation appears after a few days, the material should be prepared within a few days of its anticipated use.

SAMPLE COLLECTION AND ANALYSIS

The dithiocarbamate derivative has an exchange capacity of approximately 1 meq/g. Metal ions such as Cu^{+2}, Ag^+, Hg^{+2}, Pb^{+2}, Co^{+2}, Ni^{+2}, Zn^{+2} and Fe^{+3} are quantitatively extracted by the immobilized functional group between pH 5.5 and 9. Some metal ions such as Mn^{+2}, Ca^{+2}, Sn^{+2} are not as quantitatively extracted. Each sample type must be evaluated before the method is used routinely. A small column is prepared using 250 mg of the glass beads dispensed by a powder dispenser. Each end of the column is fitted with a polypropylene frit and an end cap for connection to flexible tubing. These columns are conveniently prepared from Econo Columns (Bio-Rad Laboratories). Samples are prepared by pumping the sample solution through the columns using a peristaltic pump. To remove solids a polypropylene frit may be used as a prefilter if desired. However, the one in the column serves this purpose. In the case of lake water, the natural pH is often within the usable range and the sample may be taken by pumping directly from the lake or stream.

X-RAY MEASUREMENTS

Once the sample is collected, the glass beads are removed from the column, mixed with an equal weight of powdered cellulose and pressed into a pellet. Pellet diameters of 12.7 mm and 37 mm are used for 100 mg and 250 mg of glass, respectively. Standards are prepared in the same manner as the samples.

X-Ray measurements were made using a Philips PW-1410 wavelength dispersive spectrometer, which is interfaced to a NOVA 1220 computer. The computer is used for goniometer control and data acquisition. This system permits rapid, sequential analysis of the sample. Counting times for each element never exceeded 100 seconds, and at least 10,000 counts were acquired for each element in a given sample.

RESULTS

The results obtained to date have been primarily on laboratory samples, although some field and industrial testing has been done. Table 11.1 shows the precision of recovery of a few cations all present at 50 μg/l in a single solution that is pumped through a column containing 250 mg of the 200-400 mesh glass beads (CPG-10) prepared according to Reaction 11.2. The flow rate of the solution was 60 ml/min and 1 liter of sample was used. Standard pellets were prepared from solutions in the concentration range 2-800μg/l. These solutions were prepared by dilution of 10^{-3} M solutions that had been standardized by titration with EDTA. The high percentage of recovery indicates a contamination problem.

Table 11.2 shows the results of recovery of ions from lake water samples. In this case, one liter of lake water was collected in a plastic bottle and a few drops of NH_3/NH_4ClO_4 buffer added. The column was prepared from the diamine-treated glass shown in Reaction 11.1 rather than the dithiocarbamate of Reaction 11.2. As will be pointed out

Table 11.1   Precision of Recovery of Metal Ions at 50 μg/l

Sample	Element (net counts/second)			
	Cu	Ni	Co	Zn
1	3734	5223	3660	4178
2	3907	5171	3653	4039
3	3686	5158	3617	3940
4	3958	4963	3554	3910
5	3569	5012	3580	3817
6	3631	5110	3596	3788
7	3597	5279	3633	4080
Mean	3712	5132	3613	3965
S.D. (%)	3.5	1.9	1.1	3.6
% Recovery	104	106	112	–

later, this material is capable of extracting certain anions as well as cations. The samples were pumped through the columns at the lake site. It should be noted that samples 3 and 4 were taken near an active marina. It is not surprising that zinc and nickel were higher there because of gasoline additives and galvanized materials.

One further point should be made. The Z-6020 (diamine) columns have been found to be very effective in removing certain anions from solution. Only oxyanions such as selenate, arsenate, molybdate, tungstate, vanadate and others appear to be so removed. An interesting application of this is the use of Z-6020 immobilized on silica to extract molybdate from the heteroacid complex phosphomolybdic acid.[6] Under the conditions used, the molybdenum so extracted is quantitatively related to the

Table 11.2   Lake Water Samples

Sample	Element Concentration (μg/l)			
	Zn	Cu	Ni	Mn
1	0.5	0.5	0.5	287
2	0.5	0.5	0.5	151
3	2.1	1.0	6.0	96
4	2.0	1.0	8.4	44
5	0.6	0.5	1.5	59

amount of orthophosphate in the sample. Based upon an X-ray determination of molybdenum, as little as 7 μg/l of phosphorus may be determined in 50 ml of ground water.

CONCLUSIONS

The use of chemically treated glass surfaces appears to offer a viable alternative for sampling dissolved ions in aqueous solutions. There are many chemical functional groups to be evaluated. However, the true test lies in adequate critical field testing.

REFERENCES

1. Carlton, D. T. and J. C. Russ. Federation of Analytical Chemistry and Spectroscopy Societies, 2nd national meeting, Indianapolis, Indiana, October, 1975.
2. Law, S. L. and W. J. Campbell. In *Advances in X-Ray Analysis*, vol. 17, C. L. Grant, C. S. Barrett, J. B. Newkirk and C. O. Ruud, Eds. (New York: Plenum Press, 1974), p. 279.
3. Leyden, D. E., T. A. Patterson and J. J. Alberts. *Anal. Chem.* 47:733 (1975).
4. Braun, T. and A. B. Farag. *Anal. Chim. Acta* 71:133 (1974).
5. Luke, C. L. *Anal. Chim. Acta* 41:237 (1968).
6. Leyden, D. E., W. K. Nonidez and P. W. Carr. *Anal. Chem.* 47:1449 (1975).

SECTION IV

CALIBRATION STANDARDS FOR X-RAY ANALYSIS

12

CALIBRATION OF ENERGY-DISPERSIVE
X-RAY SPECTROMETERS FOR ANALYSIS OF
THIN ENVIRONMENTAL SAMPLES

R. D. Giauque, R. B. Garrett and L. Y. Goda

Lawrence Berkeley Laboratory
Berkeley, California

INTRODUCTION

The ability to perform quantitative measurements with an X-ray fluorescence spectrometer is limited to the accuracy with which the output, in terms of counts/sec, can be converted to mass concentration, $\mu g/cm^2$. The task of calibrating an X-ray spectrometer involves the determination of sensitivity factors for each element of interest. In analytical programs in which 30 or more elements can be measured routinely, calibration could become a tedious and expensive problem. Fortunately, the multielement detection capabilities of energy dispersive X-ray spectrometers, together with the fact that sensitivity typically varies as a smooth function with atomic number, permit the calibration process to be greatly simplified.

In this chapter we describe several calibration techniques that can be used to standardize for analysis. The assumption is made that the specimens to be analyzed are sufficiently thin, so that matrix effects are negligible.

The first of the methods to be described for thin-film calibration requires the use of elemental thin-film standards that have been prepared by vacuum vapor deposition of individual elements onto thin substrate. The masses of the standards are determined by weighing. The second method entails the utilization of multielement standard

153

solutions that are nebulized and collected on thin substrata. The deposits are used to determine relative elemental sensitivity factors, and absolute calibration of the system is accomplished using an evaporated single element thin-film standard. The third method is a semiempirical approach that relies upon published photoelectric cross-section and fluorescence yield data to calculate relative excitation efficiencies. Detector efficiency for various X-ray lines is determined experimentally. The absolute calibration is again achieved using an evaporated single element thin-film standard.   Finally, a modification of the semiempirical approach applicable for thick standards in a few limited cases is described.   The technique can be used in cases when the thin-film criteria are not easily realized, such as for the very light elements.

CALIBRATION METHODS

To calibrate a spectrometer for analysis of thin environmental specimens it is necessary to determine the sensitivity of the system for each of the elements to be analyzed. The sensitivity, S, for an element i may be expressed:

$$I = S_i m_i \qquad (12.1)$$

where: I = the intensity, counts/sec, of the X-ray line from an elemental thin-film (matrix effects presumed to be negligible)

S_i = the sensitivity in counts/sec per $\mu g/cm^2$

m_i = the mass of the elemental thin-film, $\mu g/cm^2$.

Calibration Using Individual Elemental Thin-Film Standards

For some elements, calibration has been accomplished using individual elemental thin-film standards. The standards were prepared by vacuum vapor deposition of pure elements onto thin high purity substrata. Typical deposits made were in the range of 50-150 $\mu g/cm^2$.   Both aluminum (800 $\mu g/cm^2$ —prepared by vaporization of 99.99% Al) and Kapton (3 mg/cm^2 polyamide film) have been used as supporting substrata for the deposits. A distance of 25 cm between the vapor source and the collection substrate was maintained during the preparation of the vacuum vapor deposits.   This tended to ensure a uniform deposit over a 5-cm^2 area. The amounts of deposits collected were determined by weighing, with estimated accuracies of ± 3%.   The thin-film standards are stored in a vacuum desiccator to minimize oxidation of the deposits.

Pure elemental thin-film standards prepared at this laboratory using the vacuum vapor deposition technique include chromium, manganese, iron, nickel, copper, arsenic, selenium, silver, gold and lead. Several elements for which nonuniform deposits were made included titanium, zinc and cadmium. The lack of uniformity was determined by analyzing various areas of the deposits. Deposits made of aluminum and silicon on Kapton suffered from X-ray absorption, which was due to both loading and particle size effects. Thin-film standards prepared by vacuum vapor deposition of the elements have been successfully used for calibration by a number of laboratories.[1]   The disadvantage of this method is the cost of obtaining standards for a large number of elements.

Calibration Using Nebulized Multielement Standard Solution Deposits

Calibration for nearly all of the elements to be determined by X-ray fluorescence analysis can be accomplished by using nebulized multielement standard solution deposits on thin filters. The deposits are analyzed to determine relative elemental sensitivity factors, and absolute calibration of the system is accomplished using a single element thin-film standard prepared by vacuum vapor deposition of a pure element.

The multielement standard solutions used in this method contain from two to five elements, one of which is an internal standard element. A DeVilbiss glass nebulizer, illustrated in Figure 12.1, is employed to generate a very fine mist that is collected on a thin filter. A distance of approximately 5 cm is maintained between the nebulizer and the filter, which can be either Nuclepore polycarbonate or cellulose ester membrane.

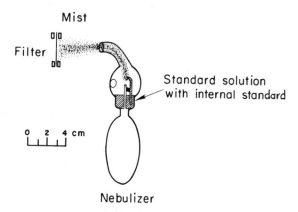

Figure 12.1   Apparatus for preparation of nebulized standard solution deposits.

Cellulose ester membrane filters should not be used for deposits of elements that have X-rays of energies less than 10 keV. A small amount of the mist can be absorbed in the filter, which in turn can attenuate some of these X-rays.

The standard solutions are prepared for most elements using a standard weighing form. For example, for many elements pure metals dissolved in acids are used. In some cases, pure compounds are dissolved in water or ammoniacal solutions (*i.e.*, $K_2Cr_2O_7$, As_2O_3, KIO_3). For a standard sulfur solution, H_2SO_4, which has been titrated with a standard NaOH solution, is used. Multielement standard solutions are prepared from aliquots of the individual element standard solutions. Mixtures are made only in combinations that will not yield overlapping X-ray lines or produce chemical reactions that will cause the concentration of an element to decrease by the formation of either a precipitate or a volatile compound. The concentration of each element in the mixed solution is in the 100 to 5000 ppm range.

Elemental deposits of between 1 to 10 $\mu g/cm^2$ are prepared and require from 50 to 500 ejections from the nebulizer. To prevent large droplet formations on the substrate, a drying period of one minute is allowed between each group of 10 ejections from the nebulizer. For elements having X-ray energies of less than 3 keV, mixtures of only two elements each are made, and the concentration of the elements in solution should not be higher than a few hundred ppm. This permits the mist droplets to evaporate to a smaller particle size and minimizes the possibility of particle size effects.

For each element, three separate standard mixed solutions of varying elemental concentration ratios are prepared (the concentration of the internal standard element usually. is not varied). Duplicate deposits are made for each standard mixed solution. Relative excitation-detection calibration factors utilized are the average values determined from the six deposits. Typically, the standard deviations (2 σ) for the relative excitation-detection efficiencies we have determined are 2% or less in nearly every case. Although the deposits are not necessarily uniform over the entire area of deposit, the use of one element as an internal standard compensates for slight nonuniformity. To minimize possible calibration errors for elements having higher energy K X-rays (>20 keV), where the detector efficiency can be influenced by nonuniformity of the deposit, the internal standard element should have X-rays within the same general energy range (± 5 keV).

To illustrate the relative ease and accuracy with which calibration can be achieved using nebulized multielement standard solution deposits, two separate NBS steel samples, 101c and 121a, were analyzed. Five hundred mg of each steel, in the form of turnings, were dissolved using a mixture of HNO_3 and HCl acids and brought up to volume. (Note: A very small amount of dark residue remained undissolved and was most likely composed of silicon and carbon, which are present at low levels in these specimens.) Three separate solutions of varying concentrations were made from each of the steel solutions and zinc was added as the internal standard element. Duplicate deposits were made for each solution on polycarbonate filters. Typical loadings corresponded to 3-10 $\mu g/cm^2$ of the original steel.

The results obtained are shown in Tables 12.1 and 12.2. The determined values are the average values obtained from the six deposits made for each steel. The deviations listed are 2σ values. Seven minutes were required to analyze each deposit. The results obtained for the major constituents—chromium, iron and nickel—agree to within 2% or better in each case with listed NBS values. Somewhat poorer agreement was obtained for manganese. However, these examples serve to illustrate the ease and capability of the technique of calibration with nebulized multi-element standard solution deposits.

Table 12.1   Analysis of NBS Steel 101c

	Determined	NBS
Cr	18.14% ± 0.38	18.21%
Mn	0.90% ± 0.08	0.640%
Fe	70.75% ± 0.54	(70.7%)[a]
Ni	9.35% ± 0.24	9.27%

[a]Fe value = 100.0% minus sum of all NBS listed constituents.

Table 12.2   Analysis of NBS Steel 121a

	Determined	NBS
Cr	18.29% ± 0.42	18.69%
Mn	1.38% ± 0.12	1.28%
Fe	68.43% ± 0.48	(68.2%)[a]
Ni	10.73% ± 0.12	10.58%

[a]Fe value = 100.0% minus sum of all NBS listed constituents.

Calibration Using a Basic Physical Approach

This method is a semiempirical approach that relies upon published photoelectric cross-section and fluorescence yield data to calculate relative excitation efficiencies. Detector efficiency for various X-ray lines is determined experimentally. Relative elemental sensitivity factors utilized combine the calculated relative excitation efficiencies with the determined detector efficiencies. The absolute calibration of the system is accomplished using an evaporated single element thin-film standard. This semiempirical approach has been described in detail elsewhere.[2] The relative elemental sensitivity calibration factors are the product of the relative probabilities of four separate processes, each of which we treat individually.

Calculation of the Relative Ability of the Excitation Radiation to Photoelectrically Produce a Vacancy in a Particular Energy Level

First considered is the probability that a photoelectric interaction will produce a vacancy in a particular inner energy level. The photoelectric mass absorption coefficient for a particular energy level may be expressed:

$$\tau \cdot (1 - \frac{1}{J_{K,L}}) \tag{12.2}$$

where: τ = the total photoelectric mass absorption coefficient, in cm^2/g, of the element for a specific energy level plus all lower energy levels

 $J_{K,L}$= the ratio (jump ratio) between the photoelectric mass absorption coefficients at the top and the bottom of the absorption edge energy.

For absorption occurring in the K shell, the value of τ is the total photoelectric mass absorption coefficient for the exciting radiation. However, for the L energy levels the value of τ is obtained by extrapolation of the curve for the particular energy level to the effective exciting radiation energy.

Calculation of the Fraction of the Vacancies Filled by Transitions that Give Rise to the Emission of a Specific X-Ray Line

Only a fraction of the vacancies created in a particular energy level are filled by transitions that give rise to the direct emission of X-rays. Some vacancies are filled by transitions involving the emission of Auger electrons. The fraction of vacancies filled by transitions that directly yield X-rays is the fluorescence yield value ($\omega_{K,L}$).

Transitions to a particular energy level give rise to the emission of more than one X-ray line since the transitions can originate from different initial energy states. The fraction of a specific X-ray line emitted with respect to the total is referred to as the fractional value (f). Hence, the net fraction of vacancies filled by transitions that give rise to the emission of a specific X-ray line may be expressed:

$$\omega_{K,L} \cdot f \tag{12.3}$$

Thus, for a particular excitation radiation, X-ray excitation curves may be established for X-ray lines from individual energy levels by multiplying the value of the terms τ, $(1 - 1/J_{K,L})$, $\omega_{K,L}$, and f. The value of each of the terms is reported in the literature.[3-7]   Figure 12.2 illustrates some calculated curves for the excitation of characteristic $K\alpha$ and $L\alpha$ X-rays with MoK radiation. As shown, the curves are a smooth function of atomic number and only several points for each curve need be plotted. Relative excitation factors for this range of elements can be interpolated to within ± 5%.

Calculation of the Fraction of X-Rays
Attenuated in the Medium Between
the Specimen and the Detector

A small proportion of the X-rays excited from the specimen may be attenuated by the air path, if present, and by the beryllium window. The fraction transmitted (T) may be expressed:

$$T = e^{-(\mu_{air} \, m_{air} + \mu_{Be} \, m_{Be})} \tag{12.4}$$

where: μ_{air} and μ_{Be} =   the total mass absorption coefficients, in cm^2/g, of air and the beryllium window for the fluorescent X-rays, respectively

m_{air} and m_{Be} =   the masses of the air and the beryllium window, in g/cm^2, that the fluorescent X-rays must transverse.

If the length of the air path is less than 2 cm and if the thickness of the beryllium window is less than 0.03 mm, the value of T would be higher than 0.97 for all X-rays of energy 6 keV and higher. Consequently, for many X-rays attenuation effects are minor or negligible.

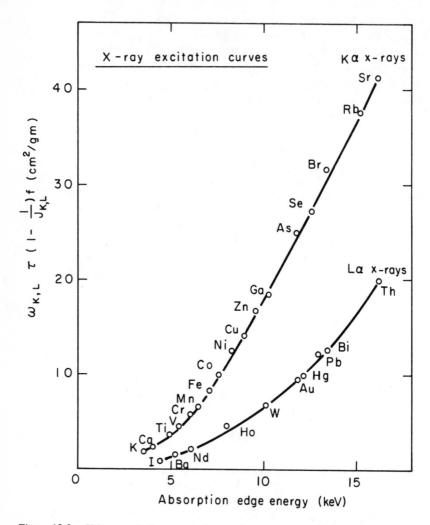

Figure 12.2   X-Ray excitation curves for excitation with molybdenum K radiation.

Determination of the Detector Efficiency
for Specific X-Ray Energies

Due to geometry and detector efficiency considerations, only a fraction of the X-rays that intersect the detector may be detected. The detector efficiency (ϵ) for radiation intersecting the detector at an angle of $90°$ near the center of the sensitive region of the detector can be calculated from the mass (m) and the photoelectric mass absorption coefficient of the detector and may be expressed:

$$\epsilon = 1 - e^{-Tm} \qquad (12.5)$$

If radiations between 4 and 15 keV strike the detector in the described manner, the efficiency is unity. However, with geometries employed for analysis, a fraction of the radiation intersects the detector at angles of less than 90° and impinges upon the detector near the periphery of the sensitive region.

To determine the effect of geometry on the detector efficiency, the intensities of characteristic X-rays from thin specimens are measured in two geometries. One is the mode employed for analysis, and the other is a highly collimated geometry, in which the X-rays detected intersect the detector in the center of the sensitive region at an angle close to 90°, as illustrated in Figure 12.3. Intensity measurements are made for

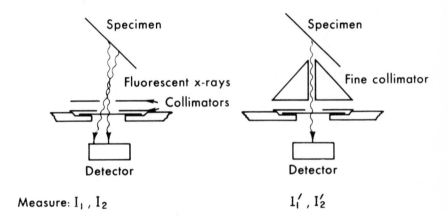

Measure: I_1 , I_2 I'_1 , I'_2

Figure 12.3 Schematic of method used to determine the detector efficiency.

two separate X-ray energies in the two geometries. The efficiency of the detector is calculated by using individual element X-ray line intensities. An X-ray line for which the detector efficiency would be unity in either geometry (*i.e.*, TiKα) is selected as a reference radiation. The detector efficiency for the second radiation may be expressed:

$$\epsilon_2 = \frac{I_2 \times I'_1}{I_1 \times I'_2} \cdot [1 - e^{-\mu_2 m}] \qquad (12.6)$$

where: I_1, I'_1, I_2, and I'_2 = the X-ray line intensities, counts/sec, for
the radiations 1 and 2 in the two geometries

μ_2 = the total mass absorption coefficient of the
detector, cm^2/g, for radiation 2

m   = the mass thickness, g/cm^2, of the detector.

Comparison of Calculated and Determined
Relative Excitation-Detection Efficiencies

Table 12.3 lists a comparison of calculated and determined relative
excitation-detection efficiencies for an X-ray system that employs a
molybdenum transmission X-ray tube.[2] The determined values are the
average values ascertained from nebulized standard solution deposits (six
for each element) and the errors listed are 2σ values. As shown, the
calculated values for eight of the elements (K, Ca, Cr, Mn, Fe, Ni, Zn
and As) normalized to the CuKα value agree to within ± 4% or better
in each case. However, the discrepancies for Rb and Sr are somewhat
larger, ± 12%. These examples serve to illustrate that relative excitation-
detection efficiencies can be calculated fairly accurately using the basic
physical approach.

Table 11.3   Relative Excitation-Detection Efficiencies

Line	Calculated	Determined
KKα	0.0622	0.0625 ± 0.0018
CaKα	0.0977	0.1005 ± 0.0026
CrKα	0.339	0.346 ± 0.006
MnKα	0.415	0.432 ± 0.006
FeKα	0.572	0.572 ± 0.010
NiKα	0.815	0.817 ± 0.006
CuKα	1.000	1.000
ZnKα	1.072	1.078 ± 0.008
AsKα	1.467	1.430 ± 0.056
RbKα	1.641	1.851 ± 0.024
SrKα	1.641	1.848 ± 0.014

Calibration Using Thick Pure Element Disks

This method is a modification of the semiempirical approach and has
been used to calibrate for analysis of very light elements for which the thin-
film criteria are not easily realized. We have used thick pure element disks
to calibrate for the analysis of light elements. This method requires that

the excitation radiation be a well collimated, near monochromatic X-ray beam. Additionally, the photoelectric cross section should be a large fraction (over 90%) of the total mass absorption coefficient for the excitation radiation. The mass, m_{thick} (g/cm^2), of the thick element disk may be expressed:

$$m_{thick} = 3.92/(\mu_e csc\psi_1 + \mu_f csc\psi_2) \qquad (12.7)$$

where: μ_e and μ_f = the total mass absorption coefficients of the element (cm^2/g) for the excitation and fluorescent radiations, respectively

ψ_1 and ψ_2 = the angles formed by the excitation and fluorescent radiations with the surface of the disk.

Since m_{thick} represents the mass for which only 25% of the radiation (excitation x fluorescence) is not attenuated, the mass of the disk for calibration purposes is $m_{thick}/4.0$.

It is important that the surface of the disk be very smooth when using this method. We have previously applied this approach to calibrate a spectrometer for analyses of three low atomic number elements—aluminum, silicon, and sulfur.[8] Calculated relative elemental sensitivity factors agreed to within 5% of the determined values, which were ascertained using thick disks for these three elements.

SUMMARY

Four separate techniques for calibrating energy dispersive X-ray spectrometers have been described. They include the use of (1) individual evaporated elemental thin-film standards, (2) nebulized multielement standard solution deposits to determine relative elemental sensitivity factors, (3) a semiempirical approach to calculate relative elemental sensitivity factors, and (4) thick pure element disks. The first three techniques are applicable for a broad range of elements. The utilization of nebulized multielement standard solution deposits, along with an evaporated single element thin-film standard for absolute system calibration, is the most accurate method of the calibration techniques described.

ACKNOWLEDGMENTS

The authors wish to thank Karl Scheu for preparing a number of vacuum vapor elemental thin-film standard deposits. We are grateful to J. M. Jaklevic for his comments on the preparation of this paper.

This work was done with support from the U.S. Energy Research and Development Administration. Any conclusion or opinions expressed in this chapter represent solely those of the author(s) and not necessarily those of Lawrence Berkeley Laboratory nor of the U.S. Energy Research and Development Administration.

REFERENCES

1. Camp, D. C. and A. L. Van Lehn. "Intercomparison of Trace Element Determinations in Simulated and Real Air Particulate Samples," *X-Ray Spect.* 4:123 (1975).
2. Giauque, R. D., F. S. Goulding, J. M. Jaklevic and R. H. Pehl. "Trace Element Determination with Semiconductor Detector X-Ray Spectrometers," *Anal. Chem.* 45:671 (1973).
3. McMaster, W. H., N. K. Del Grande, J. H. Mallett and J. H. Hubbell. *Compilation of X-Ray Cross Sections,* UCRL 50174, Sec. II and IV (1969). Available from the National Technical Information Service.
4. Bambynek, W. *et al.* "X-Ray Fluorescence Yields, Auger and Coster-Kronig Transition Probabilities," *Rev. Mod. Phys.* 44(4):716 (1972).
5. McGuire, E. J. "Atomic L-Shell Coster-Kronig, Auger and Radiative Rates and Fluorescence Yields for Na-Th," *Phys. Rev.* A3:587 (1971).
6. Hansen, J. S., H. U. Freund, and R. W. Fink. "Relative X-Ray Transition Probabilities to the K-Shell," *Nucl. Phys.* A142:604 (1970).
7. Johnson, G. G., Jr. and E. W. White. *X-Ray Emission Wavelengths and keV Tables for Nondiffractive Analysis,* ASTM DS 46, Am. Soc. Testing Matter (1970).
8. Giauque, R. D., R. B. Garrett, L. Y. Goda, J. M. Jaklevic and D. F. Malone. "Application of a Low Energy X-Ray Spectrometer to Analyses of Suspended Air Particulate Matter," in *Advances in X-Ray Analysis,* vol. 19, R. W. Gould, C. S. Barrett, J. B. Newkirk, and C. O. Ruud, Eds. (Dubuque, Iowa: Kendall/Hunt Publishing Co., 1976), pp. 305-321.

13

SOLUTION-DEPOSITED STANDARDS
USING A CAPILLARY MATRIX AND LYOPHILIZATION

R. M. Baum, R. D. Willis, R. L. Walter and W. F. Gutknecht

Duke University and Triangle Universities
Nuclear Laboratory
Duke Station, North Carolina

A. R. Stiles

Northrop Services, Inc.
Research Triangle Park, North Carolina

INTRODUCTION

A major barrier that prevents X-ray analysts from delivering quantitative results has been the problem of accurately calibrating the overall efficiency of their systems for measuring elemental abundances. In our laboratory where the mode of excitation is a proton beam of uniform intensity over a well-defined area, it is permissible to use a calibrated standard that is nonuniformly deposited on a substrate, provided the area of irradiation totally encompasses the area of the deposit. As our beam diameter is 0.4 cm, this condition is not hard to achieve if the deposit does not spread on the substrate too readily. For most X-ray fluorescence (XRF) systems, the area of irradiation is considerably larger in diameter than 0.4 cm; therefore this type of localized deposit might be considered for calibrating such systems also. However, many of these systems do not have a uniform flux, and furthermore the solid angle subtended by the X-ray detector will vary appreciably for different portions of the irradiated samples for broad beam systems, or for tight geometry arrangements. Hence, for many XRF systems it is necessary to employ standards that are uniformly deposited over an area at least as large as the area of the irradiation pattern. Such a standard would also be well-suited to the smaller diameter beams used in particle-induced X-ray emission analysis (PIXE).

Our objective in this project was to develop moderately uniform standards to be used for the calibration of X-ray fluorescence spectrometers. When the project originated, it was believed that the most useful and durable standards for calibrating broad-beam XRF systems would be ones manufactured by deposition of standard solutions onto conventional filter media, and this was the direction we pursued. Earlier experience with standards prepared employing a drop-by-drop technique led us to the conclusion that it was necessary to wet the *entire* surface uniformly and all in one step. Furthermore it was desired to develop a technique that would be simple and inexpensive, and would produce standards with a high degree of precision. Both the method we devised for this and some of the findings are described in this chapter. A report of some of the earlier results is given in Reference 1.

METHOD

The depositions on the membranes were made using a novel device that consisted of an array of 37 capillary tubes of equal, known volume mounted in an aluminum disc. Figure 13.1 displays a side view of the device. The capillaries were prepared from a piece of stock Pyrex tubing

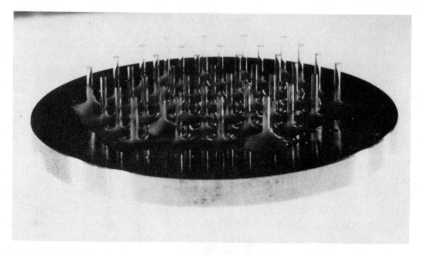

Figure 13.1 Side view of depositor exhibiting capillary matrix, aluminum disc and silicone adhesive; 37 capillaries and some extra holes can be seen.

drawn on a glass capillary drawing device. After being cut to approximately 1.2 cm in length, the capillaries were placed one by one in a die and the ends ground on fine abrasive paper so that all ends were polished and all capillaries were of equal length. The volume of each capillary was determined by weighing before and after filling with distilled water. The capillaries in this device had a mean volume of 2.44 μl with a relative standard deviation of 3.6%.

The aluminum disc was drilled with a precision milling machine to produce a regular array of holes through the disc. The capillaries were then cemented into these holes with a silicon adhesive so that one end of the capillary was flush with one side of the disc. Broken capillaries are easily replaced by removing the capillary and the adhesive around it and cementing a new capillary in place. Three such replacements were made on the device shown in Figure 13.1. Although the device appears to be fragile, it is fairly durable if handled with reasonable care. Over 1000 depositions were made with the device without any breakage of capillaries. (The three replacements were made shortly after construction because these three capillaries dried out of the plane of the rest of the capillaries.)

When dipped in a standard solution the capillaries fill by capillary action. Then the device is placed upon a filter and the capillaries drain simultaneously, wetting the entire filter. The instrument can be cleaned easily by flushing the capillaries with water. The standards produced at our laboratory were invariably free of contaminants even when solutions of different metals were deposited in succession.

In the initial work the filters were allowed to air dry after deposition of the solution. This proved unsatisfactory because with a wide variety of Millipore and Gelman filter types, migration of the metal toward the outer edge of the deposition occurred during the drying process. This produced a ring of high concentration around the deposit as illustrated schematically in Figure 13.2. Scans with PIXE across the 3-cm diagonal

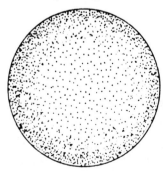

Figure 13.2 Typical density distribution of liquid deposit showing migration to the outer edges when membranes have been saturated and not freeze dried.

of these standards produced results as shown in Figure 13.3, in which the number of X-ray counts as a function of position on the standard are plotted. Each set of points is from a different single element standard. The chromium standard clearly shows the build-up of material at both edges due to the migration, and the zinc shows the trend at one edge. For chromium and several other metals, the area of the high-concentration ring constitutes a considerable portion of the total surface of the standard, thus making the areal concentration at the center of the deposit quite uncertain.

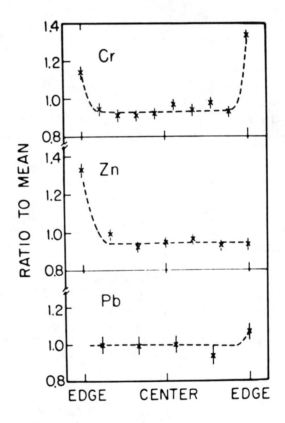

Figure 13.3 Profile of concentration obtained with PIXE scans on three mono-element nonfreeze-dried standards. Large build-up along some edges is evidenced.

Many different manipulations of the technique were explored in attempting to eliminate this problem. Different types of membrane filters from Millipore Corp. and Gelman Corp. were used. With the exception of one filter, the Gelman VM-4 which is made of vinyl chloride, all showed the same drying pattern. The Gelman VM-4 filter standards were uniform and had no build-up at the outer edge. However, the enormous amount of chlorine present in such filters created counting rate problems that affected XRF analysis for light elements. Also, when the supply of these filters on hand was exhausted, the subsequent lot of VM filters we received did not possess the same drying properties. In fact, they behaved exactly as the other types of filters that had been investigated.

From these results it was concluded that air-drying a saturated filter was unsuitable for the preparation of standards. We turned to freeze-drying the standards, and also wetting them to somewhat less than saturation. Millipore-type AAWP filters made of esters of cellulose were chosen for this stage, primarily because they are relatively free of elemental impurities. For AAWP filters, each spot from a capillary tube in the device just barely contacts adjacent spots without completely merging with them.

One special property of this filter must be considered because it significantly affects the deposition. In the manufacturing process, a stress is somehow introduced into the material. The lines of stress run parallel across the filter and can be clearly determined by visual inspection. The filters all curl slightly and the direction of the curl is perpendicular to the direction of the stress as shown in Figure 13.4. When liquid spreads

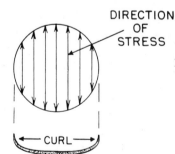

DIRECTION
OF
STRESS

CURL

Figure 13.4 Top and end views of a typical Millipore filter. Orientation of the "stress direction" relative to curl is indicated.

out from the point where a capillary contacts the filter, the flow is preferential along the lines of stress resulting in an elliptical spot from each capillary.

This flow property raises the possibility of different types of deposits depending upon the spatial relation of adjacent capillary tubes in the device with respect to the axis of the stress. In the top half of Figure 13.5, the arrangement of the capillaries in the device is represented by

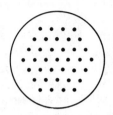

POSITION OF CAPILLARIES

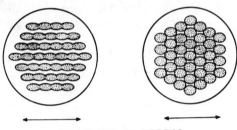

DIRECTION OF STRESS

Figure 13.5 Top half: Capillary matrix design.
Bottom half: Deposition pattern when hexagonal edge of depositor is aligned
along direction of stress (left side) and when aligned 30° relative to
direction of stress (right side).

the solid circles. The two drawings in the lower half represent *slightly
exaggerated* views of two deposition patterns that can be obtained with
the device depending upon its orientation to the direction of stress in
the filter. The drawing on the left represents the deposit that would be
obtained if the orientation of the depositor was the same as in the top
sketch. The drawing on the right would result if the depositor were
rotated 30° from the position shown at the top. In the one on the
left, the ellipses contact each other during deposition, and further flow
is restricted to within the area already wetted. On the right the capillaries are staggered along the direction of stress, and the flow is less
unidirectional. In attempting to obtain a nearly uniform deposition, the
one on the right is clearly superior.

The method chosen for freeze-drying is quite simple. Immediately
after a deposition is made, the wet standard is transferred to a clean
sheet of Teflon inside a frost-free freezer compartment. The temperature
in this freezer is between -15° and -20°C. At this temperature the standard freezes very rapidly. After approximately 30 hours the standard is
dry and can be mounted.

RESULTS

Figure 13.6 shows the results of a PIXE scan across a strontium standard prepared in this fashion. Twenty data points across a 30-mm standard were taken with a 0.4-cm diameter beam, so considerable overlap of the spots was obtained. Each point shown in Figure 13.6 represents the average of two adjacent data points except for the points at each end of the profile. The 2% error bars were calculated from counting statistics and an uncertainty to cover the estimated upper limit of a ± 1° error in the target orientation or alignment relative to the proton beam axis. This plot illustrates that a high degree of uniformity across the filter diameter was achieved with this technique. At both edges the values drop off sharply where part of the beam was positioned off the standard. There was no indication of a rise in the concentration at the outer edges as had been observed earlier; this result was anticipated since the spots barely touched one another.

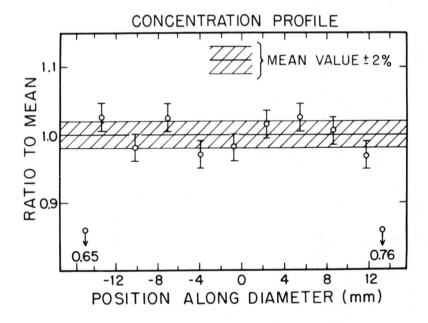

Figure 13.6 Concentration profile of freeze-dried standards showing uniformity across central region and fall-off at edges (due to partial irradiation with proton beam) without build-up in concentration as seen previously (see Figure 13.3).

The technique described here has produced standards with areal concentrations that are highly reproducible from standard to standard. In Table 13.1 the column under the heading "Liquid Deposits" and labeled "Ratio I" shows the ratio of the amount measured with an XRF system to the amount deposited for three replicates from each of several metals. For the various replicates there is no more than ± 1% deviation from the mean value for any of the metals. This is one of the most advantageous properties of the device: once a stock solution has been prepared, a large number of precise replicate standards is easily made.

Table 13.1   Results of Measurements of Replicates and Comparisons
to Thin Film Standards

| Element | Liquid Deposits | | Micromatter Foils | | Ratio I Divided by Ratio II |
	Amount Deposited ($\mu g/cm^2$)	Ratio I	Amount by Weighing ($\mu g/cm^2$)	Ratio II	
Cu	11.15	1.05	85	1.03	
	11.15	1.07	104	1.02	
	11.15	1.08	128	1.00	
		⟨1.067⟩		⟨1.016⟩	1.050
Zn	11.33	1.10	125	1.14	
	11.33	1.11			
	11.33	1.11			
		⟨1.106⟩		⟨1.14⟩	0.968
Ni	11.34	1.09	88	1.09	
	11.34	1.07			
	11.34	1.07			
		⟨1.076⟩		⟨1.09⟩	0.986
Pb	11.33	1.01	133	0.839	
	11.33	0.99	126	0.838	
	11.33	0.98			
		⟨0.993⟩		⟨0.838⟩	1.184
Mn	11.27	1.01	66	1.146	
	11.27	1.03			
	11.27	1.01			
		⟨1.017⟩		⟨1.146⟩	0.888

Note: The symbol ⟨ ⟩ signifies the average ratio.

In attempting to determine the accuracy of this technique a comparison was made between the standards we produced and standards produced by the evaporation of metal onto a thin plastic sheet. Several foil-type standards produced by the Micromatter Corp. of Seattle, Washington, had been scanned with PIXE to determine their uniformity. The ones selected for this comparison were quite uniform. The comparison was conducted with a semicalibrated, broad beam X-ray system. Table 13.1 shows the results of the comparison. The columns labeled "Ratio I" and "Ratio II" represent the ratios between the amount of metal measured experimentally with the XRF system and the amount of metal believed by their fabricators to have been deposited or evaporated. The column on the far right is the ratio between the Ratios I and II. This is a direct comparison of the two standards eliminating the calibration of the X-ray system. The values for nickel, copper and zinc agree reasonably well. However, the disagreement for manganese and lead is significant. Since the deviations are in opposite directions for manganese and lead, there does not seem to be a systematic error in the fabrication of either type of standard. It is difficult for us to find fault with the preparation of our standard solutions or with the amounts delivered to the membranes. That is, our solutions were prepared with guaranteed pure metals and the depositor, with adequate care, does deliver a reproducible amount of liquid each time. Furthermore, the PIXE scans showed that migration of the metals to the edge of the depositor standards employed did not occur. It may be significant that in the case of lead and manganese the solution deposit standards agree quite well with the calibration of the XRF system used for the analysis. Clearly there is a need to make this comparison with a suitably calibrated analytic method.

CONCLUSION

The technique outlined here has distinct promise for the preparation of membrane filter standards for X-ray analysis. The work thus far has shown that these standards are relatively uniform and are highly reproducible from standard to standard. The technique is intrinsically simple, and precise replicates can be made rapidly. Also the solution-deposited standards are expected to be durable over a long time. We feel that the technique is capable of a high degree of accuracy and is a very useful addition to the available techniques for preparation of standards.

ACKNOWLEDGMENT

The work for this chapter was supported in part by the U.S. ERDA.

REFERENCES

1. Baum, R. M., W. F. Gutknecht, R. D. Willis and R. L. Walter. *Anal. Chem.* 47:1727 (1975).

14

MASS CALIBRATION IN PIXE ANALYSES

K. Roland Akselsson

Department of Environmental Health
University of Lund
Lund, Sweden

Sven A. E. Johansson and Thomas B. Johansson

Department of Nuclear Physics
Lund Institute of Technology
Lund, Sweden

INTRODUCTION

In particle-induced X-ray emission analysis (PIXE),[1-3] one of several approaches for mass calibration may be used. For some samples it may be possible to use internal standards, thus decreasing the problems arising from the self-absorption of the X-rays and the slowing-down of the particles in the sample. For other samples back-scattered ions may be used for normalization, thus avoiding the need to weigh samples and to measure the integrated beam current.[4]

In this chapter, a simple calibration procedure is suggested, which has the advantages of both being accurate and requiring only very simple and fast checks instead of complete recalibrations. The initial calibration requires only a small number of absolutely calibrated standards since the X-ray yield function varies smoothly and slowly with X-ray energy. For the checks only a multielement standard is needed. The accelerator time needed for such a check is about three minutes and it need only be performed once every ten hours or whenever changes in the analyzing arrangement have been made. The accuracy obtained is limited mainly by the standards used, whereas the precision is determined by the counting statistics and the quality of the spectrum analysis.

175

METHOD

Small Thin Samples

PIXE is generally used for the analysis of samples smaller than the area of the proton beam. If such a sample is thin (*i.e.*, if the decrease in cross sections due to the slowing-down of the particles in the sample and the absorption of the X-rays on their way towards the detector is small), then the number of pulses Σ in a characteristic K X-ray peak is given by

$$\Sigma = n N \sigma \omega k \Omega / (4\pi) \epsilon \, T_W \, T_s \qquad (14.1)$$

n = the number of protons per cm^2

N = the number of atoms of element Z in the beam

σ = the ionization cross section

ω = the fluorescence yield

k = the relative intensity of the transitions contributing to the peak

Ω = the solid angle subtended by the detector (collimator)

ϵ = the detector efficiency

T_W = the transmission through windows (absorbers) between the sample and the detector

T_s = the average transmission in the sample.

The decrease in cross section due to the slowing-down of the particles is not taken into consideration in Equation 14.1.

Although the parameters in Equation 14.1 could be determined in principle, this would be a rather tedious operation. The errors obtained would be unnecessarily large due to difficulties in determining the efficiency and solid angle of the detector and the area of the particle beam. However, the parameters included in Equation 14.1 are either constants of smooth, slowly-varying functions of the X-ray energy or particle energy or both. There are many equivalent ways of using these features to obtain a valid mass calibration for all elements from measurements of only a small number of mass-standards.

One method of performing the mass-calibration is to determine the efficiency of the detector (including the transmission through the windows between the sample and the detector) by using a number of thin standards with X-ray energies distributed over the range of interest. Equation 14.1 written in the form

$$\epsilon T_W = \frac{\Sigma^t \, 4\pi}{n' N' \, \sigma \omega k \Omega} \qquad (14.2)$$

can then be used. The prime sign indicates that the value is obtained from measurements on standards. The "best easily obtainable" values for all the parameters are used (e.g., σ from Reference 5, ω from Reference 6 and k from References 7 and 8). Beam area and solid angle are estimated from rough geometrical measurements; their values are not crucial for the results of the PIXE-analyses, as will be discussed later.

The mass of any particular element in a thin sample ($T_s = 1$) is then calculated from the following combination of Equations 14.1 and 14.2

$$N = \frac{\Sigma \cdot 4\pi}{n\sigma\omega k\Omega} \cdot \frac{n' N' \sigma\omega k\Omega}{\Sigma' \cdot 4\pi} = \frac{\Sigma}{n} \cdot \frac{n' N'}{\Sigma'} \tag{14.3}$$

Evidently this expression is valid for elements used as standards. Since ϵT_w is a smooth function of the X-ray energy, a smooth function can then be fitted to the experimental values of ϵT_w and the results for elements not used as standards obtained.

From Equation 14.3 we see that the accuracy of Σ/Σ', n'/n and N' is critical for the accuracy of N. By using a good fitting program, Σ/Σ' is accurate to within 5%. The accuracy of n'/n obtained from commercial current integrators is much better than 5% (if moderate low-beam currents are used). The accuracy of the standards could also be better than 5%.* The homogenity of the standards is not important so long as self-absorption is low and the proton beam is uniform. As seen in Equation 14.3, errors in Ω, beam area, σ, ω and k cancel.

The accuracy of PIXE-analyses after use of the calibration procedure could be estimated from the accuracy of the standards, of the fitting program and from the goodness of the fit to the experimental values of ϵT_w. An accuracy better than 10% is easily obtained.

An advantage of this procedure and its numerous variants is that it requires very simple checks for changes in the mass calibration of the system. A standard that produces peaks in the low, medium and high energy parts of the spectrum should be mass calibrated when the system is first calibrated. This standard can then be used for future checks. Changes in the transmission between sample and detector and changes in

---

*There are commercially available ampoules with element mass known to be better than 1%. In one particular form, these contain 1.00 g of the element, which is then dissolved in 1.00 liter of water. By diluting such a solution a factor of 100, an accurate mass standard of suitable mass could be prepared by pipetting 5.00 μl on a few square millimeters of a suitable thin backing (e.g., 50 μg/cm^2 polystyrene).

the detector efficiency are reflected by the low energy peaks. A change in the detector thickness is shown by the high energy peaks. Changes in the beam area, beam current integrator calibration and detector solid angle are seen from proportional changes in all the peaks. Reasonably small changes can be corrected for in the spectrum evaluation.

It is always an advantage if no corrections are required. Thus, it is recommended that collimators in positions fixed relative to the samples be used, making the definition of the solid angle of the detector independent of small changes in the detector position. Also, external absorbers should be carefully calibrated, maintained and positioned. If the angle of a 200-μm Mylar absorber is altered by 10 degrees from being perpendicular to the sample-detector direction, the transmission of 3 keV X-rays is decreased by about 5%.

Thin Homogeneous Samples Larger than the Beam Size

PIXE is often used to analyze foils and filters in experimental arrangements that have been optimized for small samples. Equation 14.1 is valid for samples larger than the beam area if the meanings of n and N are changed to be the total number of protons and the number of atoms of element Z per cm^2 respectively. If calibration for small samples has already been performed, then only one run with a thin uniform standard larger than the beam area must be made to determine the beam area. In practice, many different standards are used to get a better determination of the beam area. The accuracy of PIXE analyses from this procedure is determined mainly by the accuracy of the uniform standards.

Of course the calibration could be carried out in a manner analogous to the small sample calibration described above by using a set of thin uniform deposit standards instead of small deposit spot standards.

In an ideal sample with uniform loading, the beam does not have to be uniform. For real samples, which may be slightly nonuniform, it is preferable to have a homogeneous beam covering as much of the sample as possible.

Thick Samples Larger than the Beam Area

Despite the introduction of uncertainties from X-ray absorption and particle deceleration, it may be feasible to use samples infinitely thick with respect to characteristic X-ray absorption and particle range. Thick samples may be used because sample preparation is made more convenient, the concentration detection limits for some elements are improved and the total mass of the sample need not be known.

The calibration procedure, however, is more difficult for thick samples. Although some work has been done using internal standards mixed into the sample, such a procedure is not always possible. If the major elements comprising the sample are fairly well-known, as is often the case with biological samples, then concentrations could be calculated from

$$\Sigma = n\,C\,\omega\,k\,\Omega/(4\pi)\,\epsilon\,T_w \int_{E_{max}}^{o} \sigma(E)\,e^{-\mu(E_x)\,x(E)}\,\frac{1}{dE/dx}\,dE \quad (14.4)$$

where: n = the total number of protons
        C = the concentration in atoms per cm^3
   $\mu(E_x)$ = the absorption coefficient
    x(E) = the depth in the sample for protons of energy E
  dE/dx = the stopping power
  E_{max} = the initial energy of the protons.

A calibration procedure similar to the one used for uniform samples could be employed but the accuracy is not as good due to additional uncertainties in the absorption coefficients, stopping power and material composition. Straggling[1,5] of the protons is another source of error, although it is of lesser importance because the increase in the cross sections from the high energy tail of the proton energy distribution at a certain depth tends to be compensated for by the decrease from the low energy tail. Also, the yields are largest at the surface where the energy spread is smallest. No extensive study on the accuracy of this routine has been done. Ahlberg and Akselsson[9] estimate the accuracy to be 10-20% for medium and high Z elements and point out the importance of the sample having a smooth surface. In thick samples with one major constituent giving a large fraction of the total X-rays in the spectrum, enhancement has to be considered for elements with absorption edges just below the intense X-ray line.

DISCUSSION

We have only discussed mass calibration from K X-rays. However, by including Coster-Kronig transitions in the model and by finding an expression for the ionization cross sections of L-shells, which differ from those of K-shells due to the different velocity distribution of their electrons, masses could be calculated from the areas of L X-ray peaks in a similar way as for the K X-ray peaks.

CONCLUSION

By using the calibration procedures suggested above, most of the systematic errors in peak area definition, beam current integration, cross sections, fluorescence yields, relative intensities, solid angle, beam area, detector efficiency and transmission factors are cancelled, and it is easy to get an accuracy better than 10% for thin samples.

REFERENCES

1. Johansson, T. B., R. E. Van Grieken, J. W. Nelson and J. W. Winchester. "Elemental Trace Analysis of Small Samples by Proton Induced X-Ray Emission," *Anal. Chem.* 47:855 (1975).
2. Walter, R. L., R. D. Willis, W. F. Gutknecht and J. M. Joyce. "Analysis of Biological, Clinical and Environmental Samples Using Proton-Induced X-Ray Emission," *Anal. Chem.* 46:843 (1974).
3. Cahill, T. A. In *New Uses of Ion Accelerators*, J. F. Ziegler, Ed. (New York: Plenum Press, 1975).
4. Lear, R. D., H. A. Van Rinsvelt and W. R. Adams. "An Investigation of the Correlation between Human Diseases and Trace Element Levels by Proton-Induced X-Ray Emission Analysis," in *Advances in X-Ray Analysis,* R. W. Gould, C. S. Barrett, J. B. Newkirk and C. O. Ruud, Eds. (Dubuque, Iowa: Kendall/Hunt Publishing Company, 1976), pp. 521-532.
5. Akselsson, R. and T. B. Johansson. "X-Ray Production by 1.5-11 MeV Protons," *Z. Physik* 266:245 (1974).
6. Bambynek, W., B. Craseman, R. W. Fink, H. U. Freund, H. Mark, C. D. Swift, R. E. Price and P. Venugopala Rao. "X-Ray Fluorescence Yields, Auger and Coster-Kronig Transition Probabilities," *Rev. Mod. Phys.* 44:716 (1972).
7. McCrary, J. H., L. V. Singman, L. H. Ziegler, L. D. Looney, C. M. Edmonds and C. E. Harris. "K-Fluorescent-X-Ray Relative-Intensity Measurements," *Phys. Rev.* A4:1745 (1971).
8. Slivinsky, V. W. and P. J. Ebert. "K_β/K_α X-Ray Transition-Probability Ratios for Elements $18 \leqslant Z \leqslant 39$," *Phys. Rev.* A5:1581 (1972).
9. Ahlberg, M. and R. Akselsson. "Proton-Induced X-Ray Emission in the Trace Analysis of Human Tooth Enamel and Dentine," *Int. J. Appl. Radiat. Isotopes* 27:279 (1976).

15

THIN LAYER STANDARDS FOR
THE CALIBRATION OF X-RAY SPECTROMETERS

R. A. Semmler, R. D. Draftz and J. Puretz

IIT Research Institute
Chicago, Illinois

INTRODUCTION

This chapter will describe a technique for making calibration standards for X-ray fluorescence analysis by depositing micron-sized particles suspended in a carrier solution onto the surface of a Nuclepore filter. The uniform deposits are typically 35 mm in diameter with clean sharp edges and masses in the range from 1 to 100 $\mu g/cm^2$. Good adhesion is achieved by depositing a thin film of collodion both before and after making the particulate deposit in order to provide both a substrate and a covering layer. Only oxides, mostly of the series IV metals, have been used thus far, but other oxides ranging from Al_2O_3 to WO_3 are planned for future single-element standards.

The essential stages in the preparation of the standards include (1) grinding and sedimentation to control particle size to 1 micron and under, (2) suspension of particles in isopropyl alcohol, (3) filtration of a large volume of suspension to gravimetrically determine concentration, (4) deposition of a collodion film on the Nuclepore substrate prior to particle deposition, (5) particle deposition, and (6) deposition of a second collodion film to complete the binding. Details of these steps and general information on the problems encountered are given in the following paragraphs.

PROCEDURE

Preparing the Suspension

Preparing suspensions involves six steps: (1) microscopic examination of pure materials, (2) grinding, (3) weighing, (4) dispersal by sonification, (5) sedimentation, and (6) reweighing.

Step 1. Powders are examined by optical microscopy to determine whether the particle size is suitable for immediate filtration or whether grinding is necessary.

Step 2. When grinding is necessary, the powders are ground in a boron carbide mortar and pestle for approximately 20 minutes and then transferred to a clean vial.

Step 3. One hundred milligrams of the ground sample is weighed in an analytical balance and transferred to a clean 150-ml beaker. Approximately 100 ml isopropyl alcohol (IPA) is added to the beaker.

Step 4. The beaker containing the suspension is placed in an ultrasonic bath for 20 minutes to break up weakly bound agglomerates. The suspension is then cooled to room temperature and the particles are allowed to settle for 20 minutes. This cycle is repeated three or four times to deagglomerate the suspension as completely as possible.

Step 5. The suspension is then allowed to settle for 20 minutes to remove the large particles. The remaining fine particles are then carefully decanted (leaving only the obvious sediment at the bottom) into a 1000-ml volumetric flask and brought to volume with IPA.

Step 6. The coarse fraction is then filtered through a preweighed Nuclepore membrane. After drying, the filter is reweighed. The difference in weight between the original sample (100 mg) and the coarse fraction gives the approximate weight of fine fraction.

Filtering

Filtering suspensions involves (1) dilution, (2) set up of filtration apparatus, (3) filtration and coating, (4) drying, and (5) mass concentration determination. The first four steps are performed at a clean bench.

Step 1. The approximate mass concentration of fine particles is used to calculate the suspension volume needed to produce the heaviest standard. The second most concentrated standard is prepared by diluting the first suspension 1:10. The least concentrated standard is prepared from 1:10 dilution of the second suspension. The standards then have a concentration ratio of 100:10:1. The same volume is used from each suspension to prepare the standards, thus ensuring uniform precision.

Step 2. The filter support and funnel are thoroughly cleaned with IPA prior to filtration. The preparation of a 35-mm diameter deposit requires an 0.8-μm Gelman membrane filter as an interface between the sintered metal support plate and the 0.4-μm Nuclepore membrane that receives the particles. The Gelman filter produces a more uniform particle deposition. The two filters are rinsed in clean IPA and carefully positioned concentrically on the filter holder, Nuclepore on top, taking care that they are perfectly flat with no sign of wrinkling. This is achieved by simultaneously wetting the filter with IPA and applying a slight vacuum to the mounted filter.

Step 3. With slight vacuum applied, the Nuclepore membrane is sprayed with a mist of 5% collodion in amyl acetate using a DeVilbiss Glass Nebulizer 40. The vacuum is gradually released and the filter funnel is carefully placed onto the mounted filter and clamped into place. Some IPA is added to the empty filter assembly; the suspension is then gradually added to the assembly. A watch glass is placed over the funnel, and vacuum is applied to the filter. During this stage of the process it is important to control the vacuum precisely, as too weak a vacuum will cause clogging of the filter and too strong a vacuum may cause nonuniform deposition, wrinkling or other problems. Occasionally, the cover glass is removed and the funnel wall gently rinsed with clean IPA to prevent sample loss. During this operation care is taken to not disturb the suspension. When the filtrate is gone, revealing the substrate, the vacuum is made weaker, the clamp is removed and the upper funnel is carefully lifted off. The filter is checked visually under a magnifying glass for uniformity and boundary sharpness. Vacuum is again applied and the filter is sprayed with a mist of 0.1% solution of collodion in amyl acetate.

Step 4. The vacuum is released and the filter is placed on a flat square of Teflon and left to dry under a large, clean glass cover.

Step 5. The exact mass concentration of fine particles is determined by filtering a large volume of the most concentrated suspension through a preweighed Nuclepore membrane. The filter and residue are dried and reweighed to determine the precise weight of fine particles delivered to each filter standard.

Comments on Potential Problems

Although powders can be purchased with a particle size specification, we have found these are often unreliable, making direct microscopic examination essential. If further grinding is necessary, a boron carbide mortar and pestle will generally be adequate and will avoid contamination with heavy elements.

Graduated cylinders proved to be easier to use and keep clean than volumetric pipettes with only a small cost in accuracy. A sintered metal support was adequate for producing a 20-mm diameter deposit but the addition of an 0.8-μm Gelman filter between the sintered metal and the Nuclepore filter was necessary to produce uniform deposits of 35-mm diameter.

If a binder is mixed directly into the filtrate for simultaneous deposition, the filter will usually clog. The filter funnel must provide a tight fit and present the same diameter opening below the filter as above in order to get streamline flow.

ERRORS AND RESULTS

The largest known source of error is introduced by the dilutions. Multiple dilution for the thinner deposits can cause an uncertainty of about 5% when using glassware with a 2% tolerance. Weighing errors affect the third or fourth significant digit and can be neglected. The deposited materials were generally spectroscopic grade chemicals with analysis data available. At least for oxides of the transition metals, particle size effects on the X-ray emission can be neglected for particles with a nominal diameter of 1 micron because self-absorption for these oxides is less than 10%. Extended comparison or application data is not yet available but initial X-ray measurements of relative response for the series IV metal oxides indicates that the response as measured by the counts/min per μg/cm^2 is constant within 18% (1σ) for deposits of a single compound that vary in total mass by two orders of magnitude.

ACKNOWLEDGMENTS

This work was supported under Contract No. 68-02-1734 with the Environmental Protection Agency, Chemistry and Physics Laboratory, Research Triangle Park, North Carolina.[1]

REFERENCES

1. Semmler, R. A. and R. G. Draftz. "Development of Standards for the Calibration of X-Ray Spectrometers for Analysis of Pollution Samples," Report No. IITRI-C6320-5 (May 1975).

SECTION V

ATTENUATION CORRECTIONS FOR PARTICLE ANALYSIS

16

ABSORPTION CORRECTIONS FOR
SUBMICRON SULFUR COLLECTED IN FILTERS

Billy W. Loo and Ray C. Gatti

Lawrence Berkeley Laboratory
University of California
Berkeley, California

Benjamin Y. H. Liu and Chong-Soong Kim

University of Minnesota
Department of Mechanical Engineering
Minneapolis, Minnesota

Thomas G. Dzubay

Environmental Protection Agency
Environmental Sciences Research Laboratory
Research Triangle Park, North Carolina

INTRODUCTION

As greater use is made of sulfur-containing fossil fuels, it will be necessary to more extensively monitor the sulfur concentration in atmospheric aerosols. Wet chemical procedures are presently used to measure sulfate extracted from the glass fiber filters used to collect particles from the atmosphere. X-Ray fluorescence (XRF) is an alternative technique that can be automated to analyze nondestructively large numbers of samples. Since no sample preparation or extraction is needed for the XRF method, a more rapid analysis at a reduced cost is possible.

For the XRF analysis of light elements such as sulfur, a potential difficulty lies in the attenuation of the characteristic X-ray by the filter medium in which the particles are collected. Such an attenuation effect is very large for glass fiber or Whatman 41 filters,[1,2] but it can be small when membrane filters are used. Previous estimates of the attenuation effect for cellulose ester filters range from 22%[3] to less than 5%.[4] In this chapter, a theoretical description of the mechanisms that cause attenuation of X-rays in air particulate samples is presented. Also presented are experimental measurements of the attenuation factor for XRF analysis of sulfur collected in cellulose ester membrane filters having 1.2-μm pore diameters. In the experiment, particles of potassium sulfate and copper sulfate are generated and collected both in the cellulose ester filter and on the surface of a 0.1-μm-pore-diameter Nuclepore filter, which is used as a reference. For particles collected in the cellulose ester filter, the attenuation factors are deduced by comparing the measured S/K and S/Cu ratios to the ratios measured for particles collected on the surface of the Nuclepore filters.

APPROACH

It is well established that urban aerosols tend to have a bimodal mass distribution. The coarse particle mode has a mass median diameter of 8 μm or larger and consists mainly of mechanically generated particles. The fine particle mode typically has a mass median diameter of 0.3 μm and consists of primary and secondary combustion products.[5,6] Previous measurements indicate that most of the sulfur occurs in the fine particle mode.[7]

To predict the attenuation effects for a set of particles having a wide range of diameters is a complex problem. The problem can be greatly simplified if the two modes of the bimodal particle distribution are collected on separate filters using a dichotomous sampler.[7-10] Then the problem consists of determining the attenuation within individual particles in the coarse fraction and of determining the attenuation in the filter and in the layer of particles for the fine fraction.[3]

Attenuation Within Individual Particles

The attenuation factor for individual particles in the shape of spheres can be approximated by[3]

$$A(\text{sphere}) = \frac{3}{2Y^3} [Y^2 - 2 + (2Y + 2)e^{-Y}] e^{-KY(\theta_1 + \theta_2)^2} \qquad (16.1)$$

where: $Y = (\mu_1 + \mu_2) D\rho$
   $K = 4 \times 10^{-6}$ (degrees)$^{-2}$ for $0 < (\theta_1 + \theta_2) < 90°$ and $0 < Y < 5$

Here μ_1 and μ_2 are the mass attenuation coefficients for incident and fluorescent X-rays within the particle, and D and ρ are the diameter and density of the particle, respectively. The angles θ_1 and θ_2 are the mean angles of incident and fluorescent X-rays as shown in Figure 16.1. This equation can describe the attenuation in a monolayer of coarse particles (particles not shadowing each other). For particles having diameters above 4 μm and having a specific gravity of 2, up to 300 μg/cm^2 of mass per unit area can be collected as a monolayer.[4] The attenuation correction calculated from Equation 16.1 can be significant for elements with atomic numbers below 18 in the coarse particle fraction. However, for sulfur in the fine particle fraction, the attenuation correction for individual particles is typically 1%.[3]

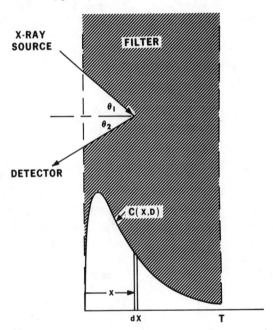

Figure 16.1 A schematic of particle deposition profile in a filter.

Layer Correction

If a uniform layer of mass is collected on the surface of a filter, the attenuation factor for the layer is given by

$$A(\text{layer}) = \frac{1 - e^{\mu L}}{\mu L} \qquad (16.2)$$

where L is the thickness (for convenience, thicknesses in this chapter are expressed in units of mass per unit area), and $\mu = \mu_1 \sec \theta_1 + \mu_2 \sec \theta_2$. For sulfur in a layer of particles with L = 200 $\mu g/cm^2$, the correction is typically about 5%.[3] Such a deposit is about the maximum loading of fine particles that a 1.2-μm-pore-diameter cellulose ester membrane filter can accept without clogging.

Filter Penetration Correction

If the particles are not collected on the surface but penetrate into the volume of the filter, then Equation 16.2 cannot be used to describe the attenuation factor. An additional absorption effect due to the passage of the X-rays through the filter medium must be included. This effect depends on the depth distribution of the particles penetrating into the filter. In this chapter, we shall determine the X-ray attenuation factor for sulfur associated with absorption in cellulose ester membrane filters (1.2-μm pore size, from Nuclepore Corporation, 7035 Commerce Circle, Pleasanton, California 94566) that are being used on a large scale in the St. Louis Regional Air Monitoring System (RAMS) network.

A schematic description of the filter penetration problem is shown in Figure 16.1. A filter of thickness T is assumed to contain a particle concentration profile C(x,D), which is a function of the depth x from the front surface of the filter and of particle diameter D. Here, the composite X-ray mass attenuation coefficient μ for the filter medium is a calculable as well as measurable quantity characteristic of the detection geometry and X-ray energies. The attenuation factor to the front side of the filter, as a function of particle size, is then

$$A(D) = \frac{\int_0^T C(x,D)e^{-\mu x}\ dx}{\int_0^T C(x,D)\ dx} \qquad (16.3)$$

On the other hand, the attenuation factor for the back side of the filter is

$$A'(D) = \frac{\int_0^T C(T-x,D)e^{-\mu x}\ dx}{\int_0^T C(T-x,D)\ dx} \qquad (16.4)$$

For ambient aerosol samples, the above expressions must be integrated over the particle size distribution W(D) so that

$$A = \frac{\int_0^\infty W(D)\ A(D)\ dD}{\int_0^\infty W(D)\ dD} \qquad (16.5)$$

It is clear that A cannot be determined without precise knowledge of C(x,D) and W(D). However, attempts have been made to deduce A by measuring the front-to-back ratio A/A' and making assumptions on C(x,D). In the present work, A(D) is determined by direct experimental measurement as described in the next section. Then a test of the frequently assumed exponential shape for C(x,D) will be described.

EXPERIMENT

Aerosol Generation and Deposition

For the purpose of determining the absorption corrections for sulfur Kα X-rays in the filters, K_2SO_4 and $CuSO_4 \cdot 5H_2O$ particles were generated in aerosol form and subsequently deposited on the filter. The aerosols were monodisperse with mass median diameters of 0.05, 0.10, 0.30 and 2.0 μm, respectively. The 2.0-μm diameter aerosol was generated by a vibrating orifice generator,[11] and the smaller particles were produced by an ultrasonic aerosol generator.

Figure 16.2 is a schematic of the ultrasonic aerosol generator used. The device consists of an ultrasonic nebulizer and an air flow system to transport, dilute, and evaporate the nebulized droplets. In addition, a radioactive Krypton 85 source of 2-mCi activity is incorporated into the system to neutralize the electrostatic charge on the particles. The ultrasonic nebulizer employed was a standard commercial device (Monoghan, 4100 East Dry Creek Road, Littleton, Colorado 80122) used for air humidification purposes. The nebulizer produces moderately monodisperse droplets with a number median diameter of 2.8 μm and a geometric standard deviation of 1.42. By nebulizing an aqueous solution of K_2SO_4 or $CuSO_4 \cdot 5H_2O$ and evaporating the solution droplets, an aerosol of the dissolved substance is produced. The final particle diameter is reduced from that of the solution droplet according to the solution concentration chosen. The 2.0-μm particles similarly produced by the vibrating orifice generator, using a 10-μm orifice, were very monodisperse, with an equivalent geometric standard deviation of about 1.05.

The aerosols generated by these methods were deposited onto 1.2-μm-pore-size cellulose ester filters under the same conditions as applied in

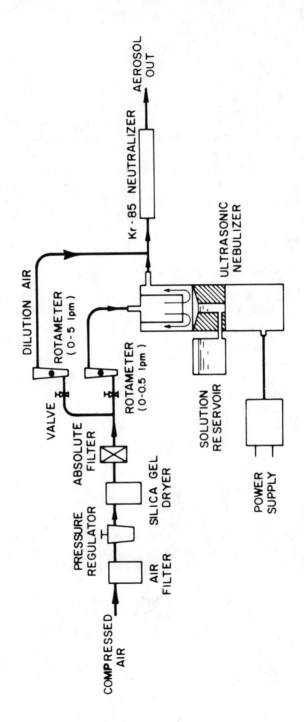

Figure 16.2 A schematic of submicron monodisperse aerosol generator.

the dichotomous samplers in the St. Louis RAMS network. With a pressure drop of approximately 20 cm Hg across the 8-cm^2 filter area, the flow rate was set to be 50 lpm at the beginning of the sampling period. Due to filter loading, the flow at the end of the sampling period dropped by up to 20%, depending on particle size and sampling time.

Similar 0.3-μm particles were also collected on Nuclepore polycarbonate filters. Since the particles were larger than the physical pore size of 0.1 μm, flow conditions were unimportant. These specimens of varying loading were prepared to serve as reference samples in which particles were known to be on the front surface.

X-Ray Fluorescence Measurements

The aerosol samples generated in the laboratory as described in the previous section were analyzed under a helium atmosphere using an energy-dispersive spectrometer with a pulsed X-ray source.[8,10] Accurate calibration was first performed for copper. The samples deposited on the surface of polycarbonate filters were then used to calibrate the system for sulfur and potassium, taking the stochiometric ratios of S/Cu and S/K to be 0.505 and 0.410, respectively. In order to minimize the uncertainties in the layer correction required, the calibration was performed by selecting a surface sample with a loading comparable to that on the test samples such that only differential layer corrections need to be applied. The copper concentrations were measured by a molybdenum secondary fluorescer (17.5 keV). Both sulfur and potassium were measured with a titanium secondary fluorescer (4.5 keV).

The measured S/K or S/Cu ratios of each test specimen relative to those of the surface reference samples immediately yielded values for A(D). In addition, the front to back ratios A(D)/A$'$(D) were determined by analyzing both sides of the filters. Other relevant quantities measured were the composite mass attenuation coefficient μ discussed earlier and the filter mass per unit area T, which was measured by a beta gauge whose accuracy had been verified by independent gravimetric determinations using a microbalance.

RESULTS

The results of the measurements of attenuation by the filter are summarized in Table 16.1. Mean values are shown together with the standard deviations for four measurements on each sample. The observed S/K and S/Cu ratios were corrected for layer attenuation effects using Equation 16.2. Since the photoelectric cross sections of S and K X-rays in K$_2$SO$_4$

Table 16.1   Results of Direct Measurements on Sulfur X-Ray Attenuation
by Filter Matrix

K_2SO_4 Particles

Filter Type[a]	Particle Diameter, D (μm)	Observed			Corrected[b]	
		S (μg/cm^2)	K (μg/cm^2)	S/K	S/K	A(D)[c]
CE	2.0	1.91	4.66	0.409 ± 0.004	0.411	1.01 ± 0.01
CE	0.3	17.48	42.84	0.408 ± 0.004	0.408	0.97 ± 0.01
CE	0.1	18.66	45.54	0.410 ± 0.004	0.410	1.00 ± 0.01
CE	0.05	19.16	49.47	0.387 ± 0.002	0.387	0.95 ± 0.01
NP	0.3	12.94	31.56	0.410 ± 0.004	0.410	1

$CuSO_4 \cdot 5H_2O$ Particles

Filter Type[a]	Particle Diameter, D (μm)	Observed			Corrected[b]	
		S (μg/cm^2)	Cu (μg/cm^2)	S/Cu	S/Cu	A(D)[c]
CE	0.41	7.98	15.93	0.501 ± 0.008	0.508	1.01 ± 0.02
CE	0.41	2.76	5.38	0.513 ± 0.009	0.505	1.00 ± 0.02
CE	0.14	22.76	5.55	0.497 ± 0.011	0.489	0.97 ± 0.02
CE	0.07	3.31	6.69	0.495 ± 0.007	0.493	0.98 ± 0.02
NP	0.41	3.88	7.69	0.505 ± 0.010	0.505	1

[a]The filters used were 1.2-μm-pore-size cellulose esters (CE) and 0.1-μm-pore-size Nuclepore (NP).

[b]Differential layer corrections were made for submicron particles and individual particle size corrections were made for the 2-μm particles.

[c]Attenuation due to filter media determined assuming A(D) = 1 for the 0.1-μm-pore-diameter Nuclepore filter.

are comparable, only small corrections were required to the raw ratios. The surface samples selected for system calibrations contained 12.94 μg/cm^2 of sulfur in the K_2SO_4 sample and 3.88 μg/cm^2 of sulfur in the $CuSO_4 \cdot 5H_2O$ sample.   Thus the differential layer corrections applied were also very small.

Figure 16.3 is a plot of the sulfur front to back ratios against thickness T of the sample filters. A general increase of A/A' with T is

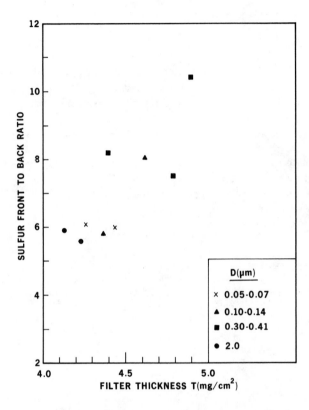

Figure 16.3   Front to back ratios vs. filter thickness.

expected. (Note: variations in the thickness T naturally occur between different filters.) The composite mass attenuation coefficient μ for titanium excitation and sulfur fluorescence X-rays was measured to be 416 ± 27 cm^2/g in the filter matrix used here.

DISCUSSION

Attenuation by the Filter

In agreement with the predictions by Giauque *et al.*,[4] the attenuation by the filter on the X-rays from sulfur is a small effect. The measurements of the attenuation by the filter shown in Table 16.1 can be summarized as

$$0.95 \leqslant A(D) \leqslant 1$$

for particle diameters in the range

$$0.05 \leqslant D \leqslant 2 \ \mu m$$

Such attenuation effects of only a few percent are considerably smaller than the 22 ± 6% effect previously predicted by Dzubay when an exponential particle deposit profile in the filter was assumed.[3]

A Test of the Exponential Deposit Profile Model

If the filter matrix appears as a large number of equally likely impaction or diffusion deposition sites along the paths of particles, then for particles of a given size the deposited concentration profile C(x) should assume the exponential form

$$C(x) = C(0)e^{-Bx} \tag{16.6}$$

Equations 16.3 and 16.4 can then be integrated to give

$$A = \frac{B}{B + \mu} \ \frac{1 - e^{-(B + \mu) T}}{1 - e^{-BT}} \tag{16.7}$$

and

$$A' = \frac{B}{B - \mu} \ \frac{e^{(B - \mu) T} - 1}{e^{BT} - 1} \tag{16.8}$$

According to the exponential model, the filter efficiency E is related to the deposit profile parameter B by

$$E = 1 - e^{-BT} \tag{16.9}$$

The efficiency of the cellulose ester filter considered in this chapter has been measured to be 99.75 ± 0.10% for 0.05 μm-particles.[12]   Thus for a filter with mass per unit area T = 4.4 x 10^{-3} g/cm^2, the values of A and A/A' should be 0.77 and 3.4, respectively. An efficiency as high as 99.9% would yield values of A and A/A' equal to 0.79 and 3.6, respectively. Since the measured values of A and A/A' were 0.95 and 7.0, the simple exponential model clearly overestimates the X-ray attenuation correction required.

SEM Observations

Extensive measurements have been made by T. Hayes of the Lawrence Berkeley Laboratory using a scanning electron microscope (SEM) equipped with an X-ray fluorescence spectrometer. Samples were coated with gold for secondary electron images and with carbon for zone scanning with XRF signals. Typical observations on 0.3-μm K_2SO_4 particles are illustrated in Figures 16.4-16.7. Figure 16.4 shows a layer of 0.3 μm K_2SO_4 particles (about 150 μg/cm^2) on the surface of a Nuclepore polycarbonate filter.

Figure 16.4 Surface of Nuclepore polycarbonate filter with 0.3-μm K_2SO_4 particles on it.

Figure 16.5 shows the back side of a cellulose ester filter. The fibers with their characteristic knobs are about 1 μm in size and are free of particles. Figure 16.6 is the front side of the same filter at the same magnification. The difference between Figures 16.5 and 16.6 in regard to the number of small (< 1 μm) features is obvious. This is due to the accumulation (about 30 μg/cm^2) of 0.3 μm K_2SO_4 particles on the filter matrix.

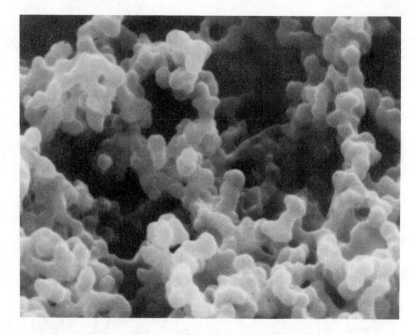

Figure 16.5   Back side of cellulose ester filter (particle-free filter matrix).

Figure 16.6   Front side of cellulose ester filter with 0.3-μm K_2SO_4 particles on it.

To determine the particle penetration profile, the same cellulose ester sample has been cooled to liquid nitrogen temperature and cracked to obtain a clean cross-sectional fracture. Figure 16.7 illustrates the five (5 μm x 5 μm) zones over which XRF signals have been integrated. Zones 1 and 2 are on the top surface (Y-Z plane) of the filter. Zone 3 spans from the surface to a depth of 2.5 μm. Zones 4 and 5 span the depths from 2.5 to 7.5 μm and 7.5 to 12.5 μm respectively. The rapid disappearance of the sulfur and potassium signals at even a few microns below the surface verifies that particles are mostly captured on or above the top layer of fiber structure in the filter.

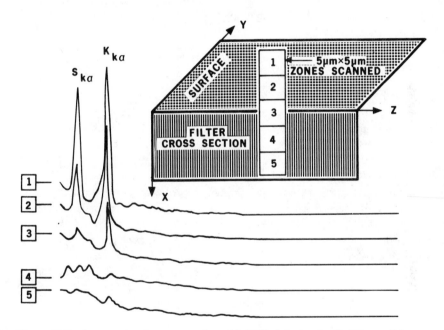

Figure 16.7 Cross-sectional zone scanning with XRF detector for K_2SO_4 particles in cellulose ester filter.

Environmental Samples Acquired in the Field

Fine particle samples from an urban (site 106) and a remote (site 124) station of the RAMS system in St. Louis have been examined for front to back ratios in the hope of determining the variability of sulfur particle size distribution that occurs in practical situations. If this distribution is sufficiently constant and narrow, then the application of attenuation

factors based on monodispersed aerosol studies to ambient samples can
be justified. Preliminary results from 60 specimens have shown no significant difference between the urban and remote samples. The front to
back ratios generally group around values consistent with surface depositions except for a few samples in which the ratio fell drastically. Such
departures were noted to occur in samples collected at the same times
at the two stations which are 40 km apart, suggesting a weather-related
phenomenon. Visual examination of the anomalous samples revealed
obvious signs of water stains, which probably caused the sulfur to migrate
into the volume of the filter. Assuming that this is the result of sampling
at high relative humidities, it should be possible to eliminate the problem
by slightly heating the incoming air stream to a temperature above the
dew point.

CONCLUSION

With the knowledge that ambient sulfur particles are predominantly
submicron, we have treated the problem of sulfur analysis by removing
the large particle interference through the use of dichotomous samplers.
After allowing for a layer correction, we have ascertained the effect of
filter matrix attenuation.

Direct X-ray attenuation measurements on $K_2 SO_4$ and $CuSO_4 \cdot 5H_2 O$
monodisperse aerosols collected on 1.2-μm cellulose ester filters show that
filter penetration is small and is not a strong function of particle size.
For ambient sulfur or sulfur-carrying particles in the range of 0.05 μm
to 1.0 μm, the filter penetration correction to X-ray measurement can be
simply represented by 3 ± 3%.

All evidence examined indicates that a reasonable model for the particle deposition profile could be of the form of a surface component
plus a component in the filter, which perhaps has an exponential depth
dependence.

$$C(x) = \begin{cases} C_o & , x = 0 \\ E\ e^{-Bx}, & x > 0 \end{cases} \qquad (16.10)$$

The relative magnitude of the two components is expected to be a function of particle size, face velocity, loading, and the structure of the filter
matrix. For the cases studied, the surface component is seen to largely
dominate over the depth component even for fairly light loadings
($\sim$ 20 μg/cm^2).

ACKNOWLEDGMENTS

We would like to express our appreciation to other staff members at the Lawrence Berkeley Laboratory who have contributed towards the success of this study, especially to R. D. Giauque for his generous collaboration, to J. M. Jaklevic, F. S. Goulding and W. L. Searles for consultations and assistance, and to T. L. Hayes for giving generously of his time and enthusiasm in performing the SEM measurements.

This report was supported in part from the United States Environmental Protection Agency under Interagency Agreement with the United States Energy Research and Development Administration. Any conclusions or opinions expressed in this report represent solely those of the authors and not necessarily those of The Regents of the University of California, the Lawrence Berkeley Laboratory, United States Energy Research and Development Administration, or the United States Environmental Protection Agency. Mention of commercial products does not constitute endorsement by any of the agencies involved.

REFERENCES

1. Adams, F. C. and R. E. Van Grieken. "Absorption Correction for X-Ray Fluorescence Analysis of Aerosol Loaded Filters," *Anal. Chem.* 47(11):1767-1733 (1975).
2. Kemp, K. "Matrix Absorption Corrections for PIXE of Urban Aerosols Sampled on Whatman 41 Filters," Chapter 17 of this volume.
3. Dzubay, T. G. and R. O. Nelson. "Self Absorption Corrections for X-Ray Fluorescence Analysis of Aerosols," in *Advances in X-Ray Analysis,* W. L. Pickles, Ed. (New York: Plenum Press, 1975), Vol. 18, pp. 619-631.
4. Giauque, R. D., R. B. Garrett, L. Y. Goda, J. M. Jaklevic and D. F. Malone. "Application of a Low-Energy X-Ray Spectrometer to Analyses of Suspended Air Particulate Matter," in *Advances in X-Ray Analysis,* R. W. Gould, C. S. Barrett, J. B. Newkirk and C. O. Ruud, Eds. (New York: Plenum Press, 1976), Vol. 19, pp. 305-321.
5. Whitby, K. T., R. B. Husar and B. Y. H. Liu. "The Aerosol Size Distribution of Los Angeles Smog," in *Aerosols and Atmospheric Chemistry,* G. M. Hidy, Ed. (New York: Academic Press, 1972), pp. 237-264.
6. Whitby, K. T. "On the Multimodal Nature of Atmopsheric Aerosol Size Distribution," Particle Technology Lab, Report No. 218 (Minneapolis: University of Minnesota, 1973).
7. Dzubay, T. G. and R. K. Stevens. "Application of the Dichotomous Sampler to the Characterization of Ambient Aerosols," Chapter 7 of this volume.

8. Jaklevic, J. M., B. W. Loo and F. S. Goulding. "Photon-Induced X-Ray Fluorescence Analysis Using Energy Dispersive Detector and Dichotomous Sampler," Chapter 1 of this volume.

9. Loo, B. W., J. M. Jaklevic and F. S. Goulding. "Dichotomous Virtual Impactors for Large-Scale Monitoring of Airborne Particulate Matter," in *Fine Particles: Aerosol Generation, Measurement, Sampling and Analysis*, B. Y. H. Liu, Ed. (New York: Academic Press, 1976), pp. 311-350.

10. Goulding, F. S., J. M. Jaklevic and B. W. Loo. "Fabrication of Aerosol Monitoring System for Determining Mass and Composition as a Function of Time," Environmental Protection Agency Report No. EPA-650/2-75-048 (Research Triangle Park, North Carolina: U.S. Environmental Protection Agency, 1975).

11. Berglund, R. N. and B. Y. H. Liu. "Generation of Monodisperse Aerosol Standards," *Environ. Sci. Technol.* 7:147-153 (1973).

12. Liu, B. Y. H. and G. A. Kuhlmey. "Efficiency of Air Sampling Filter Media," Chapter 8 of this volume.

17

MATRIX ABSORPTION CORRECTIONS FOR PIXE ANALYSIS OF URBAN AEROSOLS SAMPLED ON WHATMAN 41 FILTERS

K. Kemp

Aerosol Sciences Laboratory
Health Physics Department
Research Establishment
Riso, Denmark

INTRODUCTION

For particle-induced X-ray emission (PIXE) analysis of aerosols it is usually advantageous to use collection materials with a surface onto which the dust is deposited. Then no corrections will be necessary for absorption in such a substrate. However, in some cases the use of fiber filters is required. In these cases there can be significant depositions of the aerosol onto the volume of the filter. This chapter describes a series of measurements performed to determine the background matrix corrections for PIXE analysis of urban aerosols sampled on Whatman 41 filters (cellulose fiber). The variations of the corrections due to changes in particle size distributions or changes in humidity, which limit the precision of the analysis, are determined.

SAMPLES

It was of vital importance to use real air samples in order to determine both the corrections and the variations in actual cases. The problem incurred was to produce a set of exposed filters that could be considered representative. It is our experience from earlier measurements that the concentrations of individual elements can vary by more than a factor of

10 from day to day, but the concentrations averaged over a period of a few weeks are constant.[1]   Furthermore, the site of the sampling station within an urban area is of little importance to the daily variations.

Therefore we performed approximately 20 sets of 24-hour high volume samplings under varying meteorological conditions outside our laboratory, which is located in Copenhagen.   Two filters placed close together were exposed simultaneously.   One was a Whatman 41 filter, while the other, a Sartorius 11302 filter (cellulose nitrate, pore size 3 μm), was used as reference.   In a few cases a Sartorius 11302 filter was used as backing for the Whatman filter in order to determine the efficiency of the latter. The air velocities through the filters varied from 7 to 30 cm/s.   The inlets of the sampling tubes pointed downwards, which gave an upper cut-off particle diameter of between 40 and 100 μm due to gravitational fall-out.

ANALYSIS AND RESULTS

The Whatman filters were bombarded with a beam of 2 MeV protons, making an angle $\psi_1 = 30°$ with the normal to the filters.   Additional measurements were made for $\psi_1 = 50°$.   The X-rays emitted at an angle of 90 degrees to the incoming protons were detected in a Si(Li) detector (Figure 17.1).   The filters were bombarded both on the front and on

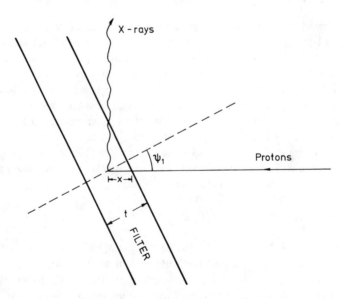

Figure 17.1   Schematic drawing showing the experimental geometry.

the rear sides. The rear side measurement gave information on the penetration of the particle into the filters. Only the results of measurements
with ψ_1 = 30° will be listed. The measurements with ψ_1 = 50° gave
similar results after taking into account the different relation between
the proton and X-ray path lengths in the filters. The Sartorius filters
were bombarded only on the front side with ψ_1 = 30°.

The ratio between the X-ray yields from the Sartorius and the Whatman
filters can be used to determine the background matrix correction factors
for the Whatman filters, if it is assumed that no corrections are necessary
as a result of penetration of dust into the membrane filters. This is
confirmed both by previous measurements of the front and rear side
yields, as described here for the Whatman filter, and by the results found
by Loo et al.[2] The mean values and uncertainties of the results and the
calculated deviations are found in Table 17.1. By using the corrections
listed, the amounts can be determined with an uncertainty of 10-15% in
addition to the analytical uncertainties.

Table 17.1   Mean Values of Ratios Between Yields from Simultaneously Exposed
Sartorius 11302 and Whatman 41 Filters for ψ_1 = 30° [a]

Element	Mean		Calculated Deviation
	Ratio	Uncertainty	
Si	2.2	10%	11%
S	3.1	10%	17%
Cl	1.1	13%	57%
K	1.8	10%	15%
Ca	1.3	10%	20%
Ti	1.4	12%	14%
V	1.8	25%	17%
Mn	1.6	13%	13%
Fe	1.5	10%	10%
Ni	1.7	30%	38%
Cu	1.3	17%	20%
Zn	1.7	10%	15%
Pb	2.0	9%	10%

[a]The uncertainties are the mean values of the analytical uncertainties. The values
in the last column are the calculated standard deviations. Calculated on the basis
of 22 sets of filters.

The ratio between the front and rear yields is shown in Table 17.2
for some selected elements. The deviations are much greater than those
in Table 17.1 because the rear yields are sensitive to the particle distributions

Table 17.2   Mean Values of Ratios Between X-Ray Yields from Whatman 41 Filters, Bombarded with Protons on the Front and Rear Side, $\psi_1 = 30°$

Element	Mean Ratio	Mean Uncertainty	Calculated Deviations	Whatman 41 Collection Efficiency	Calculated Deviation
Si	>35	–	–		
S	7.1	18%	64%	0.95	3%
K	8.0	18%	47%		
Fe	8.9	10%	18%	0.98	1%
Zn	5.9	15%	30%		
Pb	4.0	15%	23%	0.90	3%

in the filters. Table 17.2 also lists the Whatman filter efficiencies as determined by the amount of aerosol collected on the Sartorius back-up filters.

DISCUSSION

In evaluating the results, it is necessary to take into account the mass absorption coefficients μ in the filter material. The attenuation of the yields due to X-ray absorption and energy loss of the protons is calculated under the assumption that (1) the Whatman filters are solid disks with a thickness, t = 9.1 mg/cm^2, (2) the differential energy loss of the protons dE/dx is proportional to $E_p^{-\frac{1}{2}}$, (3) E_p is the proton energy, and (4) the X-ray production cross section is proportional to E_p^3:

$$C = \int_0^{t/\cos \psi_1} E_p^{-3} (x) \cdot \exp(x \cdot \mu/\tan(\psi_1)) \cdot D(x)dx$$

where D(x) is the distribution function of the particles vs. depth in the filter (see Figure 17.1).

As a first-order approximation it was assumed that D(x) = exp(-x/p). The results of calculations with p = 3.9 and 2.0 mg/cm^2 (corresponding to filter efficiencies of 90 and 99%) are shown in Table 17.3. A comparison of the results in columns 4 and 5 and the results in Table 17.2 for the front/rear side yield ratios shows clear particle-size effects. Sulfur and lead, which are known to be mainly attached to particles smaller than 1 μm, penetrate deeper into the filters than silicon and iron that are found on particles with a mean size of ~ 10 μm. Even if the trends are similar for the correction factors in columns 2 and 3 compared to

Table 17.3   Calculated Values of Matrix Correction Factors and Ratios Between
X-Ray Yields from Whatman 41 Filters, Bombarded with Protons
from the Front and the Rear Sides[a]

Element	Correction Factor		Front/Rear Side Yield	
	p=3.9 mg/cm^2	p=2.0 mg/cm^2	p=3.9 mg/cm^2	p=2.0 mg/cm^2
Si	6.7	4.2	7.7	53
S	4.0	2.7	5.6	30
K	2.6	2.0	4.0	14
Fe	2.0	1.5	3.0	7.9
Zn	1.9	1.5	2.9	7.4
Pb	1.9	1.5	2.9	7.4

[a]A dust distribution proportional to exp(-x/p) is assumed.

the values in Table 17.1, the agreement between the numbers is poor.
This indicates that the approximation to an exponential distribution in
a solid disk is too rough. To illustrate this, the front/rear side ratios
for sulfur are plotted against the Sartorius/Whatman ratios for each set
of filters together with the calculated curves for exponentially and linearly
decreasing concentrations of dust down through the filters (Figure 17.2).

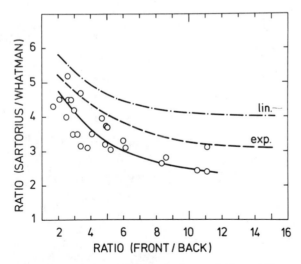

Figure 17.2 Ratios of the X-ray yields from the front and the rear side measurements on
Whatman 41 filters plotted vs. the ratio between the yields from Sartorius and Whatman
filters exposed in parallel. The solid line is a smooth curve through the data points.
In addition, curves are shown that represent calculated values for linearly and
exponentially decreasing dust concentrations down through the filter.

(The linear distribution is suggested in Reference 3.) It is seen that the actual distributions have a higher concentration of particles near the surface of the filters than the assumed distributions.

From the comparison of Whatman and membrane filters, it is concluded that absorption factors determined as the mean values of the results are correct within ± 15%. If each analysis is supplemented by a rear side measurement, the corrections can be determined with even more accuracy (≤ 10%) (Figure 17.2).

REFERENCES

1. Flyger, H., F. Palmgren Jensen and K. Kemp. *Air Pollution in Copenhagen*, Part 1, Element Analysis and Size Distribution of Aerosols, Riso Report No. 338 (1976).
2. Loo, B. W., R. C. Gatti, B. Y. H. Liu, C. S. Kim and T. G. Dzubay. "Absorption Corrections for Submicron Sulfur Collected in Filters," Chapter 16 of this volume.
3. Brosset, C. and Å. Åkerström, *Atmos. Environ.* 6:661 (1972).

18

MONTE CARLO APPLICATIONS TO THE
X-RAY FLUORESCENCE ANALYSIS OF AEROSOL SAMPLES

Alan R. Hawthorne and Robin P. Gardner

Department of Nuclear Engineering
North Carolina State University
Raleigh, North Carolina

INTRODUCTION

The use of energy-dispersive X-ray fluorescence (EDXRF) analysis as an important tool in trace element analysis of environmental samples has increased rapidly over the past few years. The advantages of EDXRF are that one can analyze for many elements simultaneously, one can process many samples in a short time, and in many cases, one can accomplish these things with very little sample preparation. The solid-state Si(Li) detectors employed in EDXRF analysis make possible the use of relatively low intensity monoenergetic isotope sources (Fe-55, Cd-109) or secondary fluorescers in conjunction with conventional X-ray tubes. These EDXRF systems often consist of an annular disc or ring source mounted coaxially about a Si(Li) detector with the sample to be analyzed placed above the source-detector configuration. It is this geometrical configuration with its wide range of sample incidence and exit angles that is particularly amenable to the Monte Carlo method for the simulation of characteristic X-ray production, attenuation and detection.

Aerosols and other environmental samples collected on filters are often sufficiently thin so that the characteristic X-rays of the analyzed elements are not significantly attenuated by the matrix containing the measured element. When there is no attenuation in the unknown sample either by the sample matrix or by the filter itself (for the case of particles small enough to penetrate into the filter), the count rate R_u for the unknown sample relative to the count rate R_s for a calibration standard is given by:

$$R_u/R_s = M_u/M_s \qquad (18.1)$$

Here M_u and M_s are the masses per unit area of the measured element in the unknown and the calibration standard, respectively. If, on the other hand, the characteristic X-rays being measured are attenuated in the unknown sample, Equation 18.1 must be modified by an attenuation factor A to account for sample attenuation.

$$R_u/R_s = A \; M_u/M_s \qquad (18.2)$$

The Monte Carlo model developed permits the calculation of attenuation factors that can be used for the correction of the measured response to the unknown sample to yield the proper elemental mass.

This chapter presents attenuation factors calculated for various practical cases for a wide-angle EDXRF system. The variation of attenuation with particle shape and size for coarse particles ($> 1 \; \mu$m) is given. The effect of enhancement on the measured response is also given for some particular cases of interest in the transition range between thin and thick samples. Attenuation of the measured intensities of fine particles ($< 1 \; \mu$m) distributed within a cellulose filter is also discussed, and attenuation factors are presented for various elements. In addition to calculating attenuation and enhancement for wide-angle EDXRF systems, the Monte Carlo model can also be used to obtain response-weighted mean entrance and exit angles for these systems. These, in turn, can be used in conjunction with analytical fixed-angle models.

MONTE CARLO MODEL FOR WIDE-ANGLE EDXRF SYSTEMS

The EDXRF system simulated in this work consists of an Fe-55 disc source mounted coaxially about a Si(Li) detector as shown in Figure 18.1. It is obvious that a wide range of photon entrance and exit angles relative to the sample normal exists in this system. The XRF process can be simulated by selecting, from the proper probability distribution functions, the various sequential events required for a source emission to result in a successful characteristic X-ray detection. Details of the model and relevant derivations are given in Reference 1. Attenuation factors can be obtained by simply comparing the intensity predicted including attenuation to the intensity predicted without attenuation. Attenuation

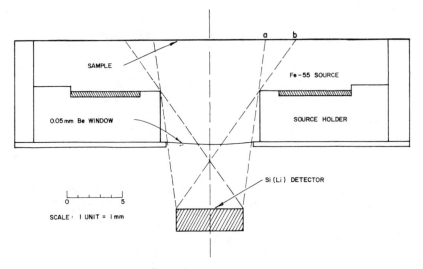

Figure 18.1 Geometry of XRF system treated in the Monte Carlo calculations.

factors are calculated for various sample shapes such as layers, infinitely long circular cylinders, and spheres for various sample matrices. In the Monte Carlo calculations the orientation of the cylinders relative to the point of intersection of the exciting X-ray and the sample is considered to be random.

The Monte Carlo model is also capable of treating the enhancement of the measured element from the characteristic X-rays of coexistent elements in the matrix. Due to the nature of the Monte Carlo model this effect is handled as easily for a thin layer sample or a monolayer of spheres as it is for a sample that may be considered infinitely thick relative to the mean-free-path of the measured characteristic X-rays.

Another application of the model is in the determination of response-weighted mean entrance and exit angles for a wide-angle EDXRF system such as the one shown in Figure 18.1. Response-weighted mean angles are determined by Equations 18.3 and 18.4 for a total number of N simulation histories.

$$\overline{\theta'} = \sum_{i=1}^{N} I_i \theta'_i / \sum_{i=1}^{N} I_i \tag{18.3}$$

$$\overline{\theta''} = \sum_{i=1}^{N} I_i \theta''_i / \sum_{i=1}^{N} I_i \tag{18.4}$$

I_i is the product of sequential probabilities required for the successful detection of the ith history, and θ'_i and θ''_i are the incident and exit angles relative to the sample normal of the ith history. These response-weighted angles can be used in various fixed-angle analytical models for the simulation of systems with a distribution of incident and exit angles. This approach has been found particularly useful in the correction for inter-element effects in Ni-Fe-Cr ternary alloys.[2,3]

Attenuation

Attenuation of the measured characteristic X-rays is most significant in XRF aerosol analysis for the K X-rays of low atomic number elements or for the L and M X-rays of higher atomic number elements. There are two particular cases of practical interest where attenuation can be a problem. The first case occurs when the particles collected on the surface of a filter are thick enough to attenuate the X-rays emerging from within the sample. This can occur in a monolayer of coarse particles (> 1 μm) or, if the sample is heavily loaded, in a layer of fine particles (< 1 μm) on the surface. The second case occurs when the fine particles are smaller than the pore size of the filter. The particles can then become embedded in the filter and the measured X-rays will be attenuated by the filter itself. These two cases may be considered separately since the attenuation within a particle small enough to penetrate into the filter is usually quite small.

One problem encountered in attempting to correct for attenuation is the requirement of knowing the distribution of collected mass. The particle size distribution is needed for coarse particles, and the depth profile distribution is needed for the fine particles within the filter. Various studies[4-6] show that coarse particles are distributed in size approximately according to a log-normal distribution. The log-normal distribution is given by:

$$\frac{dV}{dD} = \frac{1}{\sqrt{2\pi}\ \ln \sigma_g} \exp - \left[\frac{\ln (D/D_0)}{\sqrt{2}\ \ln \sigma_g} \right]^2 \tag{18.5}$$

where: D_0 = mean diameter of the distribution
σ_g = geometrical standard deviation of the distribution.

For fibrous filters, an exponential mass distribution of collected particles in the filter is usually postulated.[7] This distribution is given by:

$$\frac{dV}{dx} \propto \exp \{ x \ \ln P \} \tag{18.6}$$

where: x = the fraction of filter thickness $(0 < x < 1)$
        P = the fraction of mass penetrating through the filter.

These two distributions were used to determine the attenuation factors that are given. However, it should be pointed out that the Monte Carlo model is capable of handling any distribution, even that in discrete experimental form.

Enhancement

Although the enhancement effect is usually considered negligible,[8] there may be occasions in aerosol analysis when neglecting it results in errors greater than 10%. One case is the analysis of trace amounts of sulfur in various calcium compounds. A closed-form analytical expression[9] is available for enhancement in thick samples with fixed-angle geometry. The ratio of the sulfur K_α X-ray intensity including enhancement to the sulfur K_α X-ray intensity if enhancement is neglected is given by the following equation:

$$I'/I = 1 + \left[\mu_{S,E_1} \omega_{Ca}(1 - 1/r_{Ca})W_{Ca}\mu_{Ca,E_0}/(2\mu_{S,E_0}) \right] \cdot$$

$$\left[\left[\ell n \left(\frac{\mu_{T,E_0}/\cos\theta'}{\mu_{T,E_1}} + 1 \right) \middle/ \left(\mu_{T,E_0}/\cos\theta' \right) \right] + \right.$$

$$\left. \left[\ell n \left(\frac{\mu_{T,E_2}/\cos\theta''}{\mu_{T,E_1}} + 1 \right) \middle/ \left(\mu_{T,E_2}/\cos\theta'' \right) \right] \right] \qquad (18.7)$$

where: $\mu_{i,j}$ = mass absorption coefficient of element i at energy j
       $\mu_{T,j}$ = mass absorption coefficient of sample at energy j
       E_i = X-ray energy (subscripts: 0 = exciting energy, 1 = calcium K_α energy, 2 = sulfur K_α energy)
       ω_i = fluorescent yield of element i
       r_i = jump ratio of element i
       W_i = weight fraction of element i
       θ' = incident angle formed with sample normal
       θ'' = exit angle formed with sample normal.

However, as noted earlier, aerosol samples are generally not thick. The Monte Carlo model developed is capable of treating enhancement in the transition region between thin and thick layer samples and in spherical particles.

RESULTS

Table 18.1 shows attenuation factors for sulfur in a feldspar matrix as a function of thickness for layers, infinitely long cylinders, and spheres.

Table 18.1 S_K Attenuation Factors for a Feldspar Matrix Composed of Thin Layers, Long Cylinders Randomly Oriented, and Spheres as Calculated by the Monte Carlo Program Using the Geometry Shown in Figure 18.1

	Feldspar - $NaAlSi_3O_8$		
$d(\mu m)$	A_s (Layer)	A_s (Cylinder)	A_s (Sphere)
1	0.835	0.875	0.877
5	0.450	0.524	0.560
10	0.261	0.322	0.360
20	0.136	0.170	0.194

Note that unless attenuation is accounted for, the analysis of sulfur in feldspar spheres with a diameter of 10 μm would be underestimated by a factor of about three. The results of similar calculations are given in Reference 10 for different matrices. Table 18.2 gives attenuation factors for various elements in 10 μm spheres of two different matrices. Here the rapid increase in attenuation as X-ray energy decreases can be seen.

Table 18.2   Attenuation Factors for Various Elements in 10-μm Spheres for Two Different Matrices

	Attenuation Factors	
Element	SiO_2	$Fe_3Al_2Si_3O_{12}$
Ca	0.681	0.582
K	0.619	0.512
Cl	0.453	0.345
S	0.363	0.265
P	0.284	0.200
Si	0.519	0.187
Al	0.410	0.159

As an example of mean angle determination, Figure 18.2 shows the variation of the secant of the mean entrance angle as a function either of layer thickness or of sphere diameter for two different matrices. The fact that the mean angle is larger for a layer than for a sphere is not surprising since the path length in a layer relative to a sphere (and hence the probability of producing fluorescent X-rays) is increased for the larger entrance angles, thus weighting them more. Although the accuracy required in aerosol analysis may not warrant consideration of the variation of the mean entrance angle, similar variation encountered in thick sample ternary alloys is warranted.[3]

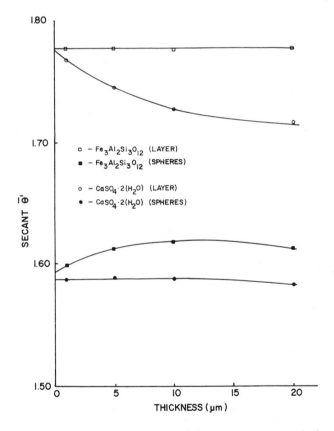

Figure 18.2   Variation of the secant of the mean entrance angle for two different matrices.

Figure 18.3 shows the effect of enhancement on sulfur analysis in various calcium compounds as a function of either layer thickness or

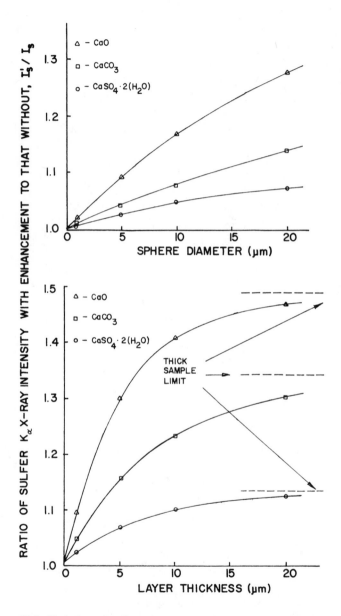

Figure 18.3 Variation of sulfur enhancement with layer thickness and sphere diameter for various calcium compounds. For CaO and CaCO$_3$ it is assumed that sulfur as a trace element is uniformly dispersed throughout the sphere or layer.

sphere diameter. Notice that the ratio of the sulfur K_α intensity-including enhancement to the intensity-neglecting enhancement approaches the thick sample limit as calculated by Equation 18.5 using Monte Carlo determined values for the mean entrance and exit angles. The magnitude of the error that results from neglecting enhancement can be seen in Figure 18.3.

To illustrate the effect of a distribution of particle sizes on the attenuation factor, the log-normal distribution of particle sizes shown in Figure 18.4 was simulated. Table 18.3 gives a comparison between attenuation factors for the distribution of sphere sizes shown in the figure and for a discrete size of 10 μm diameter. Note that for the size range considered there is little variation in the attenuation factors for the log-normal distribution as opposed to the discrete size case.

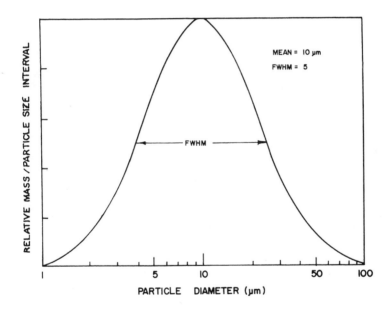

Figure 18.4 Typical log-normal particle size distribution.

Figure 18.5 shows different exponential distributions within a 5 mg/cm^2 cellulose filter. The different penetrations are labeled in terms of the fraction of fine particle mass within the first 15% of the filter thickness. Using these different distributions, attenuation factors due to the filter were calculated for the various elements listed in Table 18.4. This table shows the effect of the filter attenuation as a function of fine particle penetration. An example of where the X-rays from elements occurring in fine particles may be attenuated is $(NH_4)_2SO_4$, which may become

Table 18.3   Comparison of Sulfur Attenuation Factors for Discrete and
Log-Normal Distributions of Different Matrices

Matrix	Log-Normal Mean = 10 μm FWHM = 5[a]	Discrete Mean = 10 μm FWHM = 0[a]
$CaCO_3$	0.616[b]	0.609[b]
CaO	0.488[b]	0.497[b]
$Fe_3Al_2Si_3O_{12}$	0.241	0.245
$NaAlSi_3O_8$	0.343	0.360
SiO_2	0.327	0.363

[a]This is related to the σ_g in Equation 18.5 by FWHM = 2.354 σ_g
[b]Includes enhancement of S by Ca.

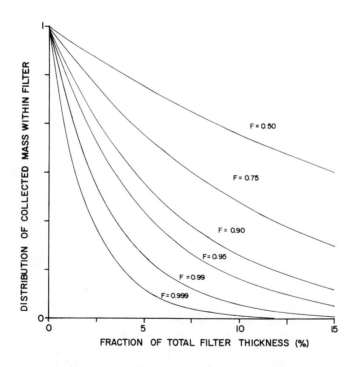

Figure 18.5   Exponential particle distribution showing fraction of mass (F)
within first 15% of filter thickness.

Table 18.4   Attenuation Factors for Fine Particles Exponentially Distributed in a 5 mg/cm^2 Cellulose Filter and Excited by an Fe-55 Source

Element	Fraction of Mass Within the First 15% of Filter					
	0.50	0.75	0.90	0.95	0.99	0.999
Ca	0.90	0.94	0.96	0.97	0.98	0.99
K	0.87	0.93	0.96	0.97	0.98	0.98
Cl	0.79	0.88	0.92	0.94	0.96	0.97
S	0.73	0.84	0.90	0.92	0.94	0.96
P	0.66	0.79	0.86	0.89	0.93	0.95
Si	0.56	0.71	0.81	0.84	0.89	0.93
Al	0.46	0.63	0.74	0.79	0.85	0.89

embedded in the filter. As $(NH_4)_2SO_4$ usually occurs in submicron particles, the attenuation within the individual particles is small, but the attenuation of the filter must be considered. Once the distribution of $(NH_4)_2SO_4$ within the filter is known, one would determine the sulfur attenuation due to the filter to obtain the required correction factor.

CONCLUSIONS

Several applications of Monte Carlo simulation to the analysis of aerosols for wide-angle EDXRF have been presented. Although the calculations are for a specific system, the general capabilities of the simulation method are well illustrated. Since Monte Carlo calculations can become quite lengthy, such simulation would not be used routinely in aerosol analysis where the samples to be analyzed are generally not easily or accurately characterized either as to their homogeneity or particle size distribution. However, use of this powerful simulation tool is justified for determination of various trends and benchmark calculations. Other treatments of the attenuation problem are given in References 11 and 12.

REFERENCES

1.   Gardner, R. P. and A. R. Hawthorne. "Monte Carlo Simulation of the X-Ray Fluorescence Excited by Discrete Energy Photons in Homogeneous Samples Including Tertiary Interelement Effects," *X-Ray Spectrometry* 4:138 (1975).

2. Hawthorne, A. R. and R. P. Gardner. "Monte Carlo Models for the Inverse Calculation of Multielement Amounts in XRF Analysis," *Trans. Amer. Nuclear Soc.* 21(Suppl. 3):38 (1975).

3. Hawthorne, A. R. and R. P. Gardner. "Fundamental Parameters Solution to the X-Ray Fluorescence Analysis of Nickel-Iron-Chromium," accepted for publication in *Analytical Chemistry*.

4. Whitby, K. T. *et al.* "The Aerosol Size Distribution of Los Angeles Smog," *J. Colloid Interface Sci.* 39:177 (1972).

5. Willeke, K. *et al.* "Size Distributions of Denver Aerosols—A Comparison of Two Sites," *Atmospheric Environment* 8:609 (1974).

6. Lundgren, D. A. "Mass Distribution of Large Atmospheric Particles," Ph.D. Thesis, University of Minnesota (1973).

7. Pich, J. "Theory of Aerosol Filtration by Fibrous and Membrane Filters," *Aerosol Sci.* 223 (1966).

8. Cooper, J. A., Ed. Workshop on X-Ray Fluorescence Analysis of Aerosols, Seattle, Wash., April, 1973. Battelle N. W. Lab. Rep. SA 4690, June, 1973.

9. Sherman, J. "The Theoretical Derivation of Fluorescent X-Ray Intensities from Mixtures," *Spectrochemica Acta* 7:283 (1955).

10. Hawthorne, A. R. and R. P. Gardner. "Monte Carlo Simulation of Self-Absorption Effects in Elemental XRF Analysis of Atmospheric Particulates Collected on Filters," *Advances X-Ray Anal.* 19:323 (1975).

11. Dzubay, T. G. and R. O. Nelson. "Self Absorption Corrections for X-Ray Fluorescence Analysis of Aerosols," *Advances X-Ray Anal.* 18:619 (1974).

12. Criss, J. W. "Particle Size and Composition Effects in X-Ray Fluorescence Analysis of Pollution Samples," *Anal. Chem.* 48:179 (1976).

19

PARTICLE SIZE EFFECTS IN POLLUTION ANALYSIS
BY X-RAY FLUORESCENCE

J. W. Criss

Naval Research Laboratory
Washington, D. C.

X-Ray absorption and fluorescence effects in particles have been studied theoretically for several years at the Naval Research Laboratory. Recently, the mathematical models were used in calculating fluorescence intensity for several thousand cases of possible interest in pollution analysis. The results—for a wide variety of particle compositions, sizes, shapes and orientations, and for different analysis conditions—led to an empirical formula for correcting measured intensities. A recent publication[1] describes the method, lists many fitted parameters, and discusses the inevitable errors in analyzing real samples. This chapter is mainly a summary of that paper, which contains references to the work of others.

When particles are all on the surface of some membrane and do not shadow each other, the effects of analysis geometry, particle shape, and particle orientation are relatively minor. The major effects depend on the incident radiation and on particle compositions and sizes; ignoring those parameters can cause the mass of an element to be underestimated by 50% for particles as small as two micrometers. These major effects are handled by a simple formula for analyzed mass per unit area on the sample:

$$M = (I/S)(1 + ba)^2 \qquad (19.1)$$

The expression I/S is the usual ratio of measured intensity to calibrated sensitivity. This ratio is all one needs for the analysis of negligibly small particles, if the calibration is based on ultrathin films (or other standards for which absorption effects are either absent or already accounted for). The remaining factor in Equation 19.1 is the particle correction for the unknown; a is the particle diameter (in micrometers) and b is a parameter that depends on particle composition and incident radiation.

The empirical correction, $(1 + ba)^2$, fits the results of elaborate calculations to a degree consistent with the usual practical uncertainty in such details as particle shapes and orientations. A table[1] of b's was constructed from such fitting, for the analysis of 48 different elements in some 200 different compounds, using radiation from either a Cr- or W-target X-ray tube operated at 40 to 50 kilovolts. Formulas are provided[1] for obtaining b's in other situations. For primary X-rays from isotopes or first-fluorescers, b can be calculated by hand. For proton-excited X-ray emission, one could simply replace the absorption coefficient for the incident radiation by the analogous parameter for protons.

A rule-of-thumb for estimating b, to determine whether or not there is a serious effect in some situation, is

$$b \approx 0.00002 \ (\mu^*_1 + \mu^*_2),$$

where μ^*_1 and μ^*_2 are the linear absorption coefficients (in cm^{-1}) of the particle for the incident and emitted radiation. If there are significant effects for the particle sizes of concern, one should use b's calculated from the better formulas.[1]

Most real samples contain particles with a range of sizes and compositions, and so one might choose effective values for b and a. A better treatment is based on rewriting Equation 19.1 as $M = I/S'$, where S' is an adjusted sensitivity. For particles of a single size and composition, $S' = S/(1 + ba)^2$; for distributions of sizes and compositions, it is best to use a value of S' that has been averaged according to the mass of particles with each size and composition.

The main source of errors in making a particle correction is the inevitable uncertainty in the distribution of sizes and compositions of the particles containing the measured element. Fortunately, most elements usually measured by X-ray emission tend to occur in combination with only a few other elements (e.g., as oxides and sulfates), so that b can be estimated fairly accurately, and the main problem is in specifying the particle sizes. For oxides and sulfates of measured metals, an uncertainty of 20% in sizes will produce an uncertainty of less than about 20% in analyzed mass, when the particle correction is less than a factor of two. Exceptional problems are encountered in measurements of sulfur and chlorine, which might occur with very light elements (e.g., ammonium sulfate) as well as with very heavy elements (e.g., lead sulfate) so that even rough analysis sometimes requires accurate knowledge of particle compositions. Even when particle compositions are known well enough, it must be remembered that indirectly-determined particle sizes (*e.g.,* aerodynamic diameters) are not the same as real sizes, and the overall size distribution for a sample does not necessarily apply to those particles containing a specific element.

In spite of uncertainty in particle sizes and compositions, it is better to make an approximate correction, based on whatever information is available, than to make no correction at all. Since real samples contain particles whose characteristics are not known precisely, there seems to be no reason to use a correction more complicated than Equation 19.1.

ACKNOWLEDGMENTS

Theoretical work was supported by the Office of Naval Research during the years 1968-1975, and has been reported in part at various meetings: Fifteenth Spectroscopy Symposium of Canada (Toronto, October 1968), XV Colloquium Spectroscopicum Internationale (Madrid, May 1969), Denver X-Ray Conference (August 1970, August 1974). The calculations for pollution analysis were supported in part by the Environmental Protection Agency under Interagency Agreement D5-0344.

REFERENCE

1. Criss, J. W. "Particle Size and Composition Effects in X-Ray Fluorescence Analysis of Pollution Samples," *Anal. Chem.* 48:179 (1976).

SECTION VI

MATHEMATICAL METHODS FOR ANALYSIS OF X-RAY SPECTRA

20

THE LINEAR LEAST-SQUARES ANALYSIS OF X-RAY FLUORESCENCE SPECTRA OF AEROSOL SAMPLES USING PURE ELEMENT LIBRARY STANDARDS AND PHOTON EXCITATION

F. Arinc, L. Wielopolski and R. P. Gardner

Department of Nuclear Engineering
North Carolina State University
Raleigh, North Carolina

INTRODUCTION

The linear least-squares analysis of complex gamma-ray spectra using library standards has been utilized for some time.[1-4] This method is now used routinely in many laboratories involved in analyzing gamma-ray spectra that contain more than one gamma-ray emitting radioisotope. The method has only recently been applied to X-ray spectra.[5-7] A significant difference between the analysis of gamma-ray and X-ray spectra is the presence of backscatter radiation from the excitation radiation in the case of fluorescence X-ray spectra. This radiation does not fit the basic least-squares assumption that the sum of the individual components at each point in the spectrum must equal that of the composite sample. Therefore, a complication is introduced in the case of X-ray spectra that is not usually present in the application to gamma-ray spectra. This can be a significant problem in those cases where the exciting radiation is complex or when the sample of interest has variable characteristics that affect the character of the backscattered radiation.

In this chapter we are interested specifically in applying the least-squares method with library standards to the XRF spectra obtained from the monoenergetic photon excitation of aerosols collected on membrane

227

filters. Fortunately, if lightly loaded filters in which the aerosol sample is 5% or less of the filter paper weight are of interest, then the back-scattered exciting radiation in the X-ray spectrum does not appreciably change in shape and varies only in total intensity.[5] This allows one to treat this portion of the X-ray spectrum as a separate library component and greatly reduces the problem of applying the method. The present chapter makes this assumption throughout.

The linear least-squares method employing library standards has the following advantages: (1) it has the most fundamental basis, (2) it is capable of giving the most accurate results possible,[4] and (3) it auto-matically provides an estimate of the accuracy of determining each individual component in the presence of all other components.

THE LEAST-SQUARES METHOD WITH LIBRARY STANDARDS

The least-squares method is based on the fundamental assumption[1,5] that the sum of the individual components at each point in a spectrum must equal that of the composite sample. If the contribution of each individual component is expressed as some constant multiplier times the spectrum of a known amount of that pure component or library spectrum, then the basic assumption for the least-squares method is that the count-ing rates in the various channels of a multichannel analyzer can be ex-pressed as

$$Y_i = \sum_{j=1}^{m} a_j R_{ij} + E_i \qquad i = 1, n \qquad (20.1)$$

where: Y_i = the counts accumulated for the fixed analysis time for the unknown composite sample in channel i

R_{ij} = the counts accumulated for the fixed analysis time for pure reference component j in channel i per unit amount of component j

a_j = the amount of component j in the unknown composite

E_i = the random error in channel i

m = the number of individual components

n = the number of channels.

To find the amounts of all individual components a_j on the basis of the maximum likelihood, one should minimize the reduced chi–square value, χ_ν^2.[8] The reduced chi-square value for ν = n-m degrees of freedom is given by

$$\chi_\nu^2 = \sum_{i=1}^{n} E_i^2 / [(n - m) \, \sigma_i^2] \qquad (20.2)$$

where σ_i^2 is the variance of E_i. If the usual assumption is made that errors in the library spectra are negligible compared to those in the spectrum of the unknown composite sample, then σ_i^2 is given approximately by

$$\sigma_i^2 \simeq Y_i \qquad\qquad i = 1, n \qquad\qquad (20.3)$$

Equation 20.2 then becomes

$$\chi_\nu^2 = \sum_{i=1}^{n} (E_i^2/Y_i) \,/\, (n - m) \qquad\qquad (20.4)$$

The values of a_k are obtained by minimization of χ_ν^2 in the usual way. A set of m simultaneous equations are solved for the a_j values that are obtained from

$$\partial\chi_\nu^2/\partial a_j = 0 \qquad\qquad j = 1, m \qquad\qquad (20.5)$$

For this set of equations, the solution in matrix notation for the vector of amounts $\underset{\sim}{a}$ with elements a_j is

$$\underset{\sim}{a} = \underset{\sim}{Z}^{-1} \cdot \underset{\sim}{r}$$

where $\underset{\sim}{Z}^{-1}$ is the inverse of the matrix $\underset{\sim}{Z}$ whose elements are given by

$$Z_{jk} = \sum_{i=1}^{n} R_{ij}R_{ik} \,/\, \sigma_i^2 \qquad j = 1, m; \quad k = 1, m \qquad (20.6)$$

and $\underset{\sim}{r}$ is the vector with elements

$$r_j = \sum_{i=1}^{n} R_{ij} \qquad\qquad j = 1, m$$

Implicit in this approach when individual library standards are to be used is that these spectra can be obtained in pure form. This is easy to accomplish for individual gamma-ray spectra but more difficult for X-ray spectra, which contain background from the X-ray source, the shielding materials, impurities in the sample substrate, and in some cases characteristic X-rays from argon in the atmosphere. These problems are solved by a method described in a later section. In the event that the standard library spectra contain errors that are not negligible compared to those of the unknown composite sample, the variance of σ_i^2 is much more complex. This case is treated in Reference 5.

ESTIMATES OF ERRORS

The standard deviations of the estimates of a_j can be obtained by using the usual Taylor series linear approximation technique. When errors in the standard library spectra can be neglected, the variances of each a_j estimate are given by

$$\sigma^2 (a_j) = \sum_{i=1}^{n} (\partial a_j / \partial Y_i)^2 \, \sigma_i^2 \quad j = 1, m \qquad (20.7)$$

Performing the indicated partial derivatives and assuming that σ_i^2 is equal to Y_i gives

$$\sigma^2 (a_j) = \sum_{i=1}^{n} \{ \text{Row } j \; [(-\underset{\sim}{Z}^{-1}) \cdot (\partial \underset{\sim}{Z} / \partial Y_i)] \cdot \underset{\sim}{a} \}^2 \, Y_i \qquad (20.8)$$

When these simplifying assumptions cannot be made, a more general technique for error evaluation can be employed, as described in Reference 5. This essentially consists of simulating the errors in Y_i and R_{ij} by choosing values of them randomly from appropriate distributions. A number of sets of these values are chosen and the a_j values are determined for each set. Then the standard deviation of each a_j is determined from the general estimator.

IMPLEMENTATION FOR AEROSOL SAMPLES ON FILTERS

The two major specific problems in the application of the linear least-squares method with library standards to X-ray spectra are: (1) eliminating the problem of the backscatter of the exciting radiation not complying with the basic assumption for the linear least-squares method and (2) obtaining the library spectra of the pure components. The first of these problems has been solved by considering the backscattered exciting radiation as a separate component. For lightly-loaded filters the backscattered radiation from monoenergetic photon sources can be assumed to vary only in total intensity from one sample to another. Therefore, a standard library spectrum for the backscattered exciting radiation can be taken as the response to a clean filter paper.[5]

The second problem is more complex. It has been solved by developing an iterative least-squares method. First, if argon is in the surrounding atmosphere, then an argon component is determined by using an atmosphere of pure argon. Then a pure backscatter library spectrum is obtained by analyzing the response to a clean membrane filter with the argon library spectrum and using the resulting residual spectrum E_i as the backscatter spectrum. This procedure is then repeated by analyzing the spectra from pure elements contained on the filter with the argon and backscatter as library spectra. The residuals are taken as the pure component library spectra.

APPLICATION PROBLEMS

As with any method of XRF spectral analysis there are problems associated with applying the linear least-squares method with complete library standards. These include: (1) statistical counting rate fluctuations that are always present to some degree, (2) overlapping characteristic X-ray peaks, (3) gain and zero shift of the analyzer system, and (4) the piling up of pulses to give false sum pulses in those instances in which high counting rates are employed. These problems are not unique to the least-squares method; in fact, the effect of these problems is reduced by this method compared to stripping or other comparable methods.[3,4] The effect of the first three of these problems was clearly shown for a realistic synthetic X-ray spectrum in Reference 5. More detailed results were given in Reference 3 for gamma-ray spectra.

To minimize the effect of statistical counting rate fluctuations one can either use more intense sources of exciting radiation or count for longer periods of time. The problem of overlapping characteristic X-ray spectral peaks can sometimes be solved by concentrating on other characteristic X-ray spectral peaks that do not overlap with those present in the sample. If the overlapping peaks do not coincide exactly, a detector system with better resolution might be employed. However, if the overlapping peaks do not coincide exactly, the least-squares method can still be used, as will be illustrated in the next section. The problem of gain and zero shift can be solved by several methods. Possibly the best solution is to stabilize the analyzer system electronically so that the problem is not encountered. If this is not feasible or desirable, one can standardize on a particular gain and zero for the analyzer system and mathematically shift all unknown and library spectra to coincide with the standard chosen before the least-squares analysis is performed.

Programs are available[3,7,9] for spectral peak location for establishing the gain and zero of a particular analyzer system and for gain shifting from the existing gain and zero to the desired ones. Another alternate scheme is to use an iterative least-squares method that adjusts the gain and zero of either the library standards or the unknown composite to obtain the minimum χ^2_ν . This procedure is described for gamma-ray spectra in Reference 3. To minimize the effect of the pulse pile-up effect one should first try to limit the total counting rates employed.

If this is not possible, one can employ electronic pile-up rejection circuits.[10] These circuits are available commercially and while they are not capable of completely eliminating the effect of pulse pile-up they are capable of reducing the spectral distortion significantly. We are also working[9,11] on mathematical methods for correcting spectra for the pulse pile-up effect.

EXAMPLE USE OF THE METHOD

Consider one of the more difficult X-ray spectral analysis problems in which chromium, manganese and iron are present with large amounts of chromium and iron and only small amounts of manganese. Note that the K_α and K_β X-rays of manganese overlap the K_β and K_α lines of chromium and iron, respectively. The three K_α X-rays have energies of 5.41, 5.90 and 6.40 keV, respectively, while the K_β X-rays have energies of 5.95, 6.49 and 7.06 keV, respectively. Two cases are simulated for the sample compositions given in Table 20.1. Actual library spectra for

Table 20.1   Example Sample Compositions for X-Ray Spectra of
Chromium, Manganese and Iron

Case	Relative Amounts Present				Counting Statistics
	Chromium	Manganese	Iron	Backscatter	
1	1.0000[a]	0.0500	1.0000	1.0000	Good
2	0.0200	0.0020	0.0200	1.0000	Poor

[a]This amount of chromium is approximately 13.7 $\mu g/cm^2$.

chromium, manganese and iron excited by a molybdenum secondary fluorescer were used in conjunction with an actual backscatter spectrum of the molybdenum fluorescer X-rays from a clean filter to produce the desired composite, synthetic spectra. Statistical counting rate fluctuations were superimposed on the synthetic spectra by choosing random numbers distributed according to a Poisson distribution. Note that the amount of manganese present is a factor of 20 or 10 lower than the amounts of chromium and iron in the two cases, and good statistics are available for the first case and poor for the second. The spectra for these two cases are shown in Figures 20.1 and 20.2.

It would be reasonable to expect that the analyst would not realize that manganese was present in either Case 1 or 2. If the linear least-squares

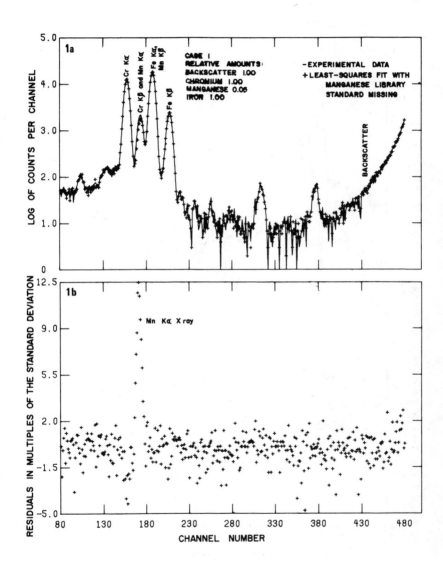

Figure 20.1 Case 1. Good Statistics. X-Ray spectra and residuals for relative amounts of backscatter, Cr, Mn and Fe in the ratios 1.00, 1.00, 0.05 and 1.00.

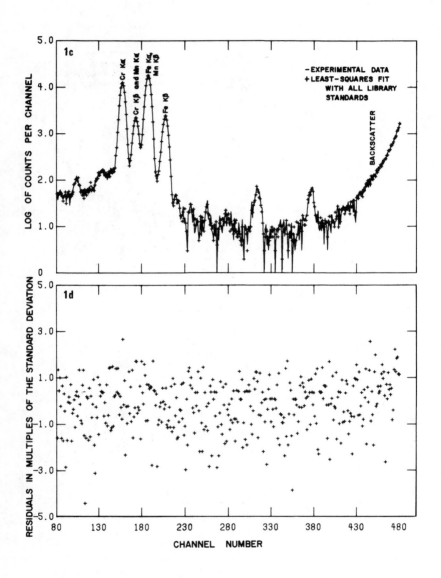

Figure 20.1, Continued

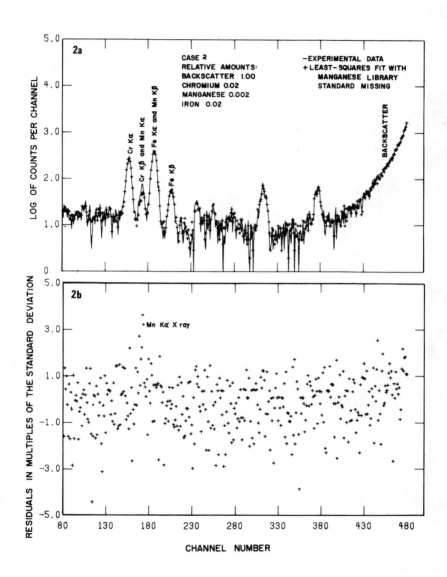

Figure 20.2 Case 2. Poor Statistics. X-Ray spectra and residuals for relative amounts of backscatter, Cr, Mn and Fe in the ratios 1.00, 0.02, 0.002 and 0.02.

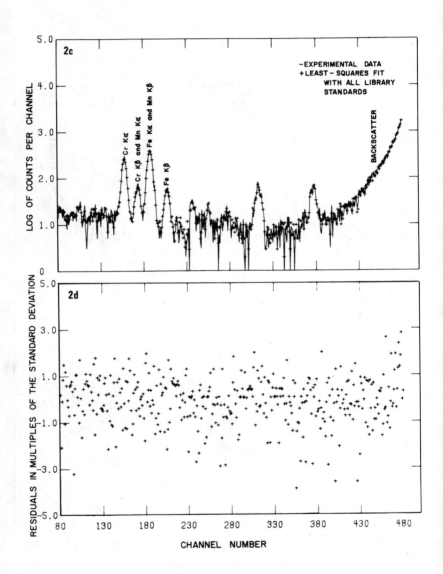

Figure 20.2, Continued

analysis was performed with only the backscatter, chromium, and iron library spectra, then the results shown in Table 20.2 and Figures 20.1a and 20.2a would be obtained. Note that the actual errors of the amounts

Table 20.2   Results of a Least-Squares Analysis of Cases 1 and 2 when only the Backscatter, Chromium and Iron Libraries are Used

| Case | Amounts Calculated by Least-Squares Method | | | Reduced Chi-Square Values |
	Chromium	Iron	Backscatter	
1	1.0315 (0.34)[a]	1.0029 (0.27)	0.9748 (0.74)	3.20
2	0.0201 (2.68)	0.0195 (2.15)	0.9718 (0.74)	1.37

[a]Numbers in the parentheses are the calculated standard deviations in percent available from the least-squares analysis.

calculated in this manner are within the predicted standard deviations for two out of the four element amounts reported. Note also that the reduced chi-square values for the least-squares fits are significantly higher than unity, indicating that there is something amiss. These high reduced chi-square values might lead the analyst to examine the residuals from the least-squares calculation. These are shown for the two cases in Figures 20.1b and 20.2b. Note that in both cases (particularly in Case 1, which has good statistics) the manganese K_α X-ray clearly appears. This would indicate to the analyst that he should have included the manganese library in the least-squares analysis. When this is done the results are as given in Table 20.3 and Figures 20.1c and 20.2c. Note that the errors in this case are within the predicted standard deviations for four out of

Table 20.3 Results of a Least-Squares Analysis of Cases 1 and 2 when the Backscatter, Chromium, Manganese and Iron Libraries are Used

| Case | Amounts Calculated by Least-Squares Method | | | | Reduced Chi-Square Value |
	Chromium	Manganese	Iron	Backscatter	
1	1.0043 (0.36)[a]	0.04819 (3.46)	0.9973 (0.28)	0.9740 (0.74)	1.10
2	0.01949 (2.80)	0.00166 (18.93)	0.01936 (2.17)	0.9707 (0.74)	1.30

[a]Numbers in the parentheses are the calculated standard deviations in percent available from the least-squares analysis.

the six element amounts reported as should be the case. Note also that the reduced chi-square values for the least-squares fits are closer to unity, indicating that all is well.

DISCUSSION AND CONCLUSIONS

The results presented in the previous section are indicative of what can be accomplished with the least-squares method under severe conditions of spectral peak overlap in the presence of variable amounts of statistical counting rate fluctuations. A number of trends can be identified from these results. First, the reduced chi-square value (χ_ν^2) is an accurate indicator of elements that have been missed if those elements represent a relatively large fraction of the total spectrum. Second, the predicted standard deviations for each elemental amount are properly indicative of the errors involved in the estimates if the controlling source of error is the statistical fluctuation of the counting rates (if χ_ν^2 is about unity). Third and finally, the residuals of the least-squares analysis are a very sensitive indicator of missed components if one could quantify the visual effect indicated in the plots of the residuals shown in Figures 20.1 and 20.2. A suggested alternative to the visual analysis of the residuals is to first use all possible component libraries in the calculation. Those libraries that give amounts that are either negative or less than some specified amount times the calculated standard deviation (say 1.00 times the calculated standard deviation) are discarded and the calculation is done again.

From the results given in the previous section it may be concluded that the least-squares method with complete library spectra is an excellent method for X-ray spectral analysis. The only major disadvantage of this method is the necessity for measuring and storing the library spectra for each element of interest. This requires considerable initial experimental effort and considerable computer storage space. We are presently working on an alternate approach that consists of determining the response function of the analyzer system for all energies of interest. This approach would allow the generalization of all library spectra with an empirical model, and it might be accomplished by a technique similar to that used by Heath et al.[12] for the generalization of gamma-ray response functions.

FUTURE WORK

We are presently working on two refinements to the least-squares method other than the complete response function approach mentioned in the previous section. One of these is the development of a model for the backscatter portion of the X-ray spectrum. For heavily-loaded filters or other samples of variable composition and thickness, the backscatter portion of the X-ray spectrum cannot be considered constant. A Monte Carlo method[13] has been developed to simulate the backscatter from homogeneous samples. The other refinement is the development of a pulse pile-up model[11] and the subsequent incorporation of this model into an iterative least-squares method. This will minimize the effect of pulse pile up in the analysis of X-ray spectra taken at high counting rates.

ACKNOWLEDGMENT

The authors gratefully acknowledge the support of this work by the U.S. Environmental Protection Agency under U.S.E.P.A. Grant No. R-802759.

REFERENCES

1. Salmon, L. "Analysis of Gamma-Ray Scintillation Spectra by the Method of Least-Squares," *Nucl. Instr. Methods* 14:193 (1961).
2. O'Kelley, G. D., Ed. Proceedings of the Symposium on the Application of Computers to Nuclear and Radiochemistry (Gatlinburg, Tennessee, October, 1962), USAEC Monograph NAS-NS 3107 (March 1963).
3. Schonfeld, E., A. H. Kibbey and W. Davis, Jr. "Determination of Nuclide Concentrations in Solutions Containing Low Levels of Radioactivity by Least-Squares Resolution of the Gamma-Ray Spectra," *Nucl. Instrum. Methods* 45:1 (1966).
4. Quittner, P. *Gamma-Ray Spectroscopy with Particular Reference to Detector and Computer Evaluation Techniques* (London: Adam Hilger, Ltd., 1972).
5. Arinc, F., R. P. Gardner, L. Wielopolski and A. R. Stiles. "Application of the Least-Squares Method to the Analysis of XRF Spectral Intensities from Atmospheric Particulates Collected on Filters," *Advan. X-Ray Anal.* 19:367 (1976).
6. Cooper, J. A., Ed. "Review of a Workshop on X-Ray Fluorescence Analysis of Aerosols," BNWL-SA-4690, Seattle, Washington (April 1973).
7. Trombka, J. I. and R. L. Schmadebeck. "A Numerical Least-Square Method for Resolving Complex Pulse-Height Spectra," NASA SP-3044 (1968).
8. Bevington, P. R. *Data Reduction and Error Analysis for the Physical Sciences* (New York: McGraw-Hill Book Co., 1969).
9. Gardner, R. P., F. Arinc, A. R. Hawthorne, L. Wielopolski, G. R. Beam and K. Verghese. "Mathematical Techniques for X-Ray Analyzers," Technical Progress Report for U.S.E.P.A. Grant No. R-802759 for the period from May 15, 1975 to May 14, 1976.
10. Goulding, F. S. and J. M. Jaklevic. "X-Ray Fluorescence Spectrometer for Airborne Particulate Monitoring," EPA Report No. EPA-R2-73-182 (April, 1973).

11.   Wielopolski, L. and R. P. Gardner. "Prediction of the Pulse-Height Spectral Distortion Caused by the Peak Pile-Up Effect," *Nucl. Instrum. Methods* 133:303 (1976).

12.   Heath, R. L., R. G. Helmer, L. A. Schmittroth and G. A. Cazier. "The Calculation of Gamma-Ray Shape for NaI Scintillation Spectrometer, Computer Program and Experimental Problems," IDO-17917 (April, 1965).

13.   Arinc, F. and R. P. Gardner. "Models for Correcting Backscatter Nonlinearities in XRF Pulse-Height Spectra," *Transactions Amer. Nucl. Soc.* 21 (Suppl. 3):37 (1975).

21

A MODIFICATION OF THE LINEAR LEAST-SQUARES FITTING METHOD WHICH PROVIDES CONTINUUM SUPPRESSION

F. H. Schamber

Tracor Northern, Inc.
Middleton, Wisconsin

INTRODUCTION

Analysis of X-ray energy spectra obtained with silicon detectors is complicated by the numerous instances of unresolved and overlapped peaks. Conventional least-squares procedures incorporating shape models for the peaks can be employed to unfold peak overlaps. However, the continuum component present in all X-ray spectra requires special treatment.

This chapter describes a simple modification of the linear least-squares procedure, which is highly insensitive to continuum. The method consists of applying a simple digital cross-correlation function to both the unknown spectrum and to the peak reference shapes prior to fitting. This digital filter operator acts in a manner analogous to a frequency band-pass filter and strongly attenuates the low-frequency components of the spectrum (continuum) while passing the frequencies characteristic of peak structure. Thus, using this "filter-fit" method, it is not necessary to formulate an explicit continuum model nor is it necessary to provide fitting parameters for the continuum. The simplicity of this procedure makes it particularly well-suited for on-line data reduction in a minicomputer.

DERIVATION

The Linear Least-Squares Procedure

With the use of a silicon detector and a multichannel pulse-height analyzer, one can measure the energy spectrum of X-radiation emitted by a fluoresced sample. Such a spectrum typically consists of several hundred "channels" of information; the content of each channel is the number of events (counts) detected within a narrow energy interval. The basic assumption of the linear least-squares method is that, apart from random errors, each of the channels of a measured spectrum can be represented simply by summing together the channels of a particular set of constituent spectra (reference spectra) in the correct proportions; that is

$$\overline{Y}_j = \sum_{i=1}^{n} A_i R_{ij} \qquad (21.1)$$

where: $\overline{Y}_j$ = the content of channel j of an idealized model spectrum
R_{ij} = the content of channel j of reference spectrum i
A_i = the fraction of reference spectrum i that is present in the composite

Each reference spectrum represents a standard shape that is due to a particular source of X-rays. Commonly the references are measured pure-element spectra.

The least-squares method provides a procedure by which a set of "best-fit" A_i coefficients may be determined for a particular measured spectrum; they are the values that minimize the quantity

$$\chi^2 = \sum_j E_j^2 / \sigma_j^2 \qquad (21.2)$$

where: $E_j = Y_j - \overline{Y}_j$
Y_j = the content of channel j of a particular measured spectrum.

Therefore E_j is the error due to counting statistics and other sources of random error; σ_j is the standard deviation of the measured channel content Y_j. When the measured values Y_j are Gaussian distributed with mean $\overline{Y}_j$ and variance σ_j^2, then it can be shown that the set of solution coefficients that minimize χ^2 are also the "maximum-likelihood" solutions (i.e., the most probable values).

Treatment of the Continuum

The linear least-squares procedure, as outlined above, was originally developed for analysis of nuclear gamma-ray spectra acquired with sodium iodide scintillation detectors.[1-3]   Such spectra are qualitatively very similar to X-ray spectra acquired with silicon detectors in that observed structure is relatively complex, and peak overlaps and interferences are very common.   The success in this application recommends it for use with X-ray spectra also.[4]   However, gamma-ray spectra result from radioactive decay of the sample material, and ambient background radiation is usually the only major spectral contaminant that must be dealt with.   This component is easily measured as the no-sample background spectrum.

On the other hand, X-ray spectra are normally obtained by fluorescence excitation, and the dominant spectral contaminants are related to the excitation mode.   Charged particle excitation produces bremsstrahlung in the sample;   photon excitation contributes elastic (Rayleigh) and inelastic (Compton) scattering from the sample and adjacent material.   These processes result in the production of a relatively smooth continuum distribution that underlies the entire spectrum and contaminates all intensity measurements.   It is the complexity of dealing with this continuum component that presents the greatest difficulty in the least-squares analysis of X-ray spectra.

Several techniques have been proposed for modeling the bremsstrahlung continuum observed in electron beam microanalysis.[4-7]   Procedures for modeling the continuum generated by photon and proton excitation are described elsewhere in this volume.

A relatively detailed continuum model is necessitated by the fact that the method of least squares fitting (as developed above) results in a matching of channel counts in the unknown versus counts in the references.   Thus, an inaccuracy in the continuum model introduces inaccuracy into the fitting of the peak shapes.   This property of the least-squares procedure is particularly frustrating, since even to the untrained eye the continuum in a spectrum is qualitatively quite different from peak structure.   It would seem that there should be some manner in which the least-squares procedure could be modified so that it obtains the best fit for *structure*, rather than the best fit for amplitude.   What is desired is a means of desensitizing the least-squares fit to slowly varying shapes without sacrificing sensitivity to the more rapidly changing peak-shaped structures.

The Use of Digital Filters for Continuum Suppression

If a pulse-height spectrum is viewed as though the horizontal axis (energy) were actually a time axis, then the observed spectral structure can be loosely characterized by three "frequency" bands: (1) a low-frequency continuum, (2) intermediate-frequency peak structure,

(3) high-frequency noise due to random statistical fluctuations from channel-to-channel. Thus, the digital equivalent of an electrical "band-pass" filter might be useful as a means of extracting the peak-structure information from the spectrum.

A digital filter can be defined by a set of coefficients that are multiplied against corresponding channels of a spectrum. The sum of these products is the "filtered" value for that point in the spectrum and can be expressed as

$$FILT(Q_j) = \sum_{s=-t}^{t} f_s Q_{j+s}$$

Here Q represents the spectrum, and the f coefficients are a set of constants. In the present context we are only interested in filters that are symmetric

$$f_s = f_{-s}$$

and have zero-weight

$$\sum_{s=-t}^{t} f_s = 0$$

The rectangular filter of Figure 21.1 is one of the most elementary of such filters; the central lobe (upper width) consists of UW positive

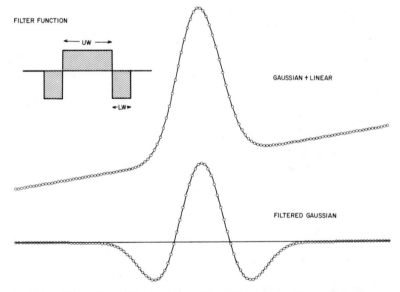

Figure 21.1 The application of a digital filter to a simple peak shape.

coefficients; the outer (lower width) lobes contain LW negative coefficients:

$$f_s = -C/(2LW), \quad -(LW+UW/2) \leqslant s < -UW/2 \qquad (21.3)$$
$$f_s = +C/UW, \quad -UW/2 \leqslant s \leqslant +UW/2$$
$$f_s = -C/(2LW), \quad +UW/2 < s \leqslant +(UW/2+LW)$$

where C is an arbitrary constant. (With C = 1, this filter is nothing more than the average of UW central channels minus the average of two regions of LW channels on either side.) When this filter is applied to the Figure 21.1 spectrum consisting of a Gaussian-shaped peak superimposed on a straight line, it results in the filtered spectrum shown immediately below. There is a strong response to the peak, and the straight line is suppressed entirely.

It is not difficult to show that any symmetric zero-weight filter has a zero-response to a straight line; *i.e.,* if $Q_j = a + bj$ over the region $j = k - t$ to $j = k + t$, then

$$FILT(Q_k) = \sum_{s=-t}^{t} f_s[a+b(k+s)] = 0$$

It then follows that any curve that is approximately linear over the width of the filter is suppressed. This is illustrated in Figure 21.2; a typical X-ray spectrum has been filtered using the rectangular filter of Figure 21.1. The continuum present in this spectrum is effectively filtered out (note the removal of the broad backscatter hump underlying the Fe lines). However, the positions of the characteristic peaks and their approximate amplitudes are still apparent in the filtered spectrum. Also note that there is moderate smoothing (suppression of high frequency components) due to the averaging effect of the filter.

Such digital filters are commonly used in spectral analysis as a means of locating and identifying peaks present on a noisy or sloping baseline.[8-11] This technique is popular because of the manner in which such functions tend to "draw out" similarly shaped spectral structure and suppress all other features.

Least-Squares Fitting Using Filtered Spectra

We consider now the application of digital filtering to the linear least-squares fitting problem. One may consider the measured spectrum to be the sum of reference spectra, statistical noise, and a "slowly-varying" continuum component.

$$Y_j = \sum_{i=1}^{n} A_i R_{ij} + E_j + cont_j \qquad (21.4)$$

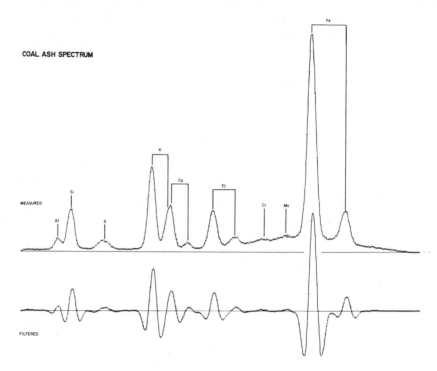

Figure 21.2   The effect of digitally filtering a complex spectrum.

It is a simple exercise to demonstrate that any digital filter is a linear operator;   that is

$$FILT(aQ_{1j} + bQ_{2j}) = a[FILT(Q_{1j})] + b[FILT(Q_{2j})]$$

Applying a symmetric, zero-weight filter (such as the one defined by Equations 21.3 to both sides of Equation 21.4 one obtains

$$FILT(Y_j) = \sum_{i=1}^{n} A_i[FILT(R_{ij})] + FILT(E_j) + FILT(cont_j)$$

Since the continuum is assumed to be a slowly varying component,

$$FILT(cont_j) \cong 0$$

and then

$$FILT(Y_j) \cong \sum_{i=1}^{n} A_i[FILT(R_{ij})] + FILT(E_j) \qquad (21.5)$$

This result is illustrated in Figure 21.3: on the left is a composite spectrum that is the sum of two Gaussian peak spectra and a smooth continuum function; on the right are the filtered composite spectrum and the two filtered peak spectra. To a very good approximation, the filtered composite spectrum is the sum of the filtered peak spectra only; no continuum component is needed.

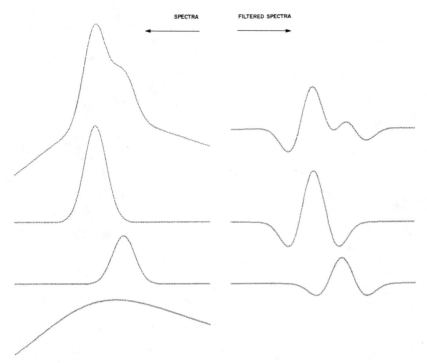

SPECTRA   FILTERED SPECTRA

Figure 21.3 Decomposition of a spectrum in unfiltered and filtered representations.

It is then reasonable to construct a least-squares fit using filtered spectra. We define a new χ^2 variable by

$$\chi^2 = \sum_j [Y_j' - \sum_{i=1}^{n} A_i R_{ij}']^2 w_j \tag{21.6}$$

with the definitions

$$Y_j' = \text{FILT}(Y_j)$$
$$R_{ij}' = \text{FILT}(R_{ij})$$
$$w_j = 1/(\sigma_j')^2$$

where $(\sigma_j)^2$ is the variance of Y_j'

$$(\sigma_j')^2 = \sum_m (\partial Y_j'/\partial Y_m)^2 \, \sigma_m^2$$

Assuming that observed errors are due to counting statistics only $(\sigma_j^2 \sim Y_j)$,

$$(\sigma_j')^2 \cong \sum_{s=-t}^{t} f_s^2 \, Y_{j+s} \tag{21.7}$$

From this point, the mathematical development employs the well-known Chi-squared minimization procedure. One proceeds by minimizing χ^2 with respect to each of the fitting coefficients

$$\partial \chi^2 / \partial A_k = 0 \qquad (k=1,n)$$

which leads to the set of coupled equations

$$\sum_j Y_j' R_{kj}' w_j = \sum_{i=1}^{n} A_i \sum_j R_{ij}' R_{kj}' w_j \qquad (k=1,n) \tag{21.8}$$

With the definitions

$$G_k = \sum_j Y_j' R_{kj}' w_j$$

and

$$Z_{ik} = \sum_j R_{ij}' R_{kj}' w_j$$

one may rewrite Equation 21.8 in the compact form

$$G_k = \sum_{i=1}^{n} A_i Z_{ik} \qquad (k=1,n) \tag{21.9}$$

The A_i coefficients are then determined by solution of the matrix equation

$$A_k = \sum_{i=1}^{n} G_k Z_{ik}^{-1} \tag{21.10}$$

where Z^{-1} is the inverse of matrix Z.

Beginning at Equation 21.8, the above is the standard linear least-squares method except that the filtered quantities Y_j' and R_{ij}' are substituted for the original data Y_j and R_{ij}, and correlated variances $(\sigma_j')^2$ are used instead of the original channel variances σ_j^2. No new parameters were necessary to account for continuum since the application of the filter prior to fitting suppresses all low frequency components. Since

Gaussian peak shapes contain low frequency components that are also suppressed by the filter, the shapes of the filtered peaks are altered considerably. However, for a well-chosen filter, the information content remaining is sufficient to permit accurate fitting.

It is instructive to note that the digital filter used in this derivation is analogous to the second-derivative operator of calculus. Thus, the method results in the transformation of the least-squares procedure from a best-fit of channel amplitude to a best-fit of curvature.

Estimate of Uncertainty

A virtue of the least-squares procedure is that it allows the uncertainties of fitted coefficients to be estimated. If the measured channel counts are independent variables and the reference spectra are known with negligible error, then the uncertainty of the A_i coefficient is obtained by the evaluation of

$$\sigma^2(A_i) = \sum_j (\partial A_i/\partial Y_j)^2 \, \sigma_j^2 \qquad (21.11)$$

In the ordinary application of the least-squares method (using unaltered channel amplitudes), Equation 21.11 reduces to the familiar result

$$\sigma^2(A_i) = Z_{ii}^{-1}$$

In the present instance, the fitting is done with respect to the correlated values Y_j' and therefore

$$\sigma^2(A_i) = \sum_j [\sum_m (\partial A_i/\partial Y_m') (\partial Y_m'/\partial Y_j)]^2 \, \sigma_j^2 \qquad (21.12)$$

A formal evaluation of Equation 21.12 is complex and would add greatly to the computational overhead. A simple approximation has been derived and empirically verified. For the rectangular operator defined by Equations 21.3 with C=1, an approximation to Equation 21.12 is given by

$$\sigma^2(A_i) = [(UW \cdot LW)/(UW + LW)] Z_{ii}^{-1} \qquad (21.13)$$

The derivation of this form assumes that the proportions of the filter approximate the shape of a spectral peak. For the filters used in this work (UW $\sim$ 160eV, LW $\sim$ 80eV) and for a spectrometer of typical resolution, this form provides a moderate overestimate of the uncertainties.

Although Equation 21.13 is only an approximation, it has been found to provide a useful estimate of statistical error.

APPLICATION

Implementation of the Filter-Fit Procedure

A major virtue of the filter-fit method is that it is a simple procedure that can be implemented in a minicomputer for on-line data reduction. To date, the filter-fit method has been incorporated in two separate programs for use in a minicomputer-based X-ray analyzer (Tracor Northern NS-880). The first program which used this method, designated ML (Multiple Least-squares),[1,2] was designed to operate in a very small minicomputer (PDP-11 with 8K words of memory) and is restricted to six elements per fit. The more recent program (SUPER-ML) uses essentially the same assembly-language subroutines to perform the filtering and generate the fitting arrays, but provides for a larger number of references and can be directly linked to data reduction programs written in a high-level language.

The execution speed of the filter-fit method is quite adequate for routine on-line usage. Unfolding of a spectrum containing six elements typically requires 10-30 seconds of computation time (depending upon energy calibration and number of channels). The major computational overhead imposed by the method is the necessity of filtering the unknown and reference spectra before use in the least-squares fit. The simple rectangular filter of Equation 21.3 is essentially the difference of two multichannel averages and can be applied quite rapidly. Both of the above computer programs filter the spectra on a channel-by-channel basis during the actual fit; this allows the original reference spectra to be retained for stripping and other uses. However, prefiltering the reference spectra could result in a substantial improvement in speed for many repetitive applications, especially if the program were to be encoded entirely in a high-level language such as FORTRAN where the filtering time is likely to be nonnegligible, or if a more highly shaped filter were to be used.

Reference Spectra

Reference spectra may be either measured or generated from mathematical formulae (*e.g.,* Gaussian peak shapes). There is only one condition that the filter-fit method imposes upon the reference spectra: they must provide an accurate model of the *peak* structure seen in the unknown.

Since both the unknown and the reference spectra are filtered before they are used in the least-squares analysis, the presence or absence of continuum in these spectra is immaterial. This greatly simplifies the preparation of suitable reference spectra since those factors that affect continuum generation (excitation voltage, substrate composition, sample thickness) can be ignored. A single multielement spectrum can also be used as the reference for several elements when the various peak groups are well separated.

No provision has been made in the fitting equations for the statistics of the reference spectra. When measured references are used, the statistical precision of the reference spectra should be substantially better than that of the unknown spectrum.

Since those portions of a reference spectrum that do not contain peaks are set to zero by the action of the digital filter, they need not be included in the fit. This permits a substantial saving in computer memory since only the peak regions of the reference spectra are needed by the fitting program. This property allows the unfolding of a spectrum to be performed piecemeal; a complex spectrum can be broken down into smaller groups of interfering elements that are unfolded together. This can be a significant program economy since the size of the fitting matrix increases as the square of the number of simultaneous fitting components.

Choice of Filter Dimensions

Since the frequency band passed by a digital filter depends strongly upon its dimensions, it is important that the filter be reasonably well matched to the spectra being analyzed. Reference 11 includes curves that relate peak detectability (signal to noise ratio) and resolution (peak broadening) to the dimensions UW and LW of a rectangular filter. These curves would indicate that optimum detectability would result for UW equal to the full width at half maximum (FWHM) of a peak and LW as large as possible. On the other hand, choosing a very large value for LW increases the sensitivity to curvature in the continuum. The filter shape used in the current computer programs has been chosen to be UW = 160 eV and LW = 60 eV, which is based upon a nominal detector resolution of FWHM $\sim$ 160 eV. At 10 eV/channel calibration, this corresponds to UW = 16, LW = 6 channels as shown in Figure 21.1. These dimensions have been found to give a reasonable compromise between sensitivity to peak shapes and rejection of continuum.

Performance Characteristics

Sensitivity to Continuum

   Figure 21.4 is a comparison of the response to a continuum component
for three different peak analysis techniques. A fixed Gaussian peak (REF)
of constant width was superimposed upon a variable-width Gaussian
(BGND) that was used to simulate a curving continuum. The amplitude
of the REF peak was then determined by each of the three methods and
the error of the determination was plotted as a function of the ratio of

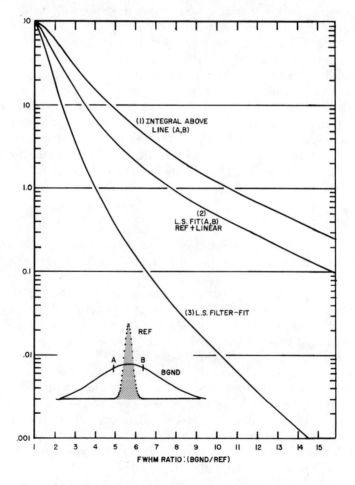

Figure 21.4   Comparison of the sensitivity of three analysis methods
to a curving continuum.

the BGND peak width to the REF peak width. The top trace was obtained by direct integration of the area above the straight line segment A,B on the BGND curve. This curve thus represents the error for the method of linear interpolation under a peak of shape REF. The middle curve was obtained by least-squares fitting over the region A,B using a Gaussian peak shape and a straight line to parameterize the background. This curve then represents the error using the ordinary least-squares method with a linear background. The bottom curve was obtained using the filter-fit method with the REF peak as the sole fitting reference (standard filter dimensions).

The linear-interpolation and fit-with-straight-line methods both drop off with increasing BGND width as expected. However, the rate at which the filter-fit method drops off is quite remarkable. Note that the filter-fit method is subject to less than 1% error when the BGND curve is only four times the width of the reference peak. This small an error is not achieved by the peak-plus-linear fit until BGND is eight times broader than REF, by which point the filter-fit method has fallen to an error of less than 0.05%. (For comparison, the width of the BGND curve shown in the figure inset is eight times that of the REF peak.)

Response to Unreferenced Peak Structure

Figure 21.5 illustrates the response of three different peak-area measurement methods to an unexpected peak. This figure was constructed by determining the amplitude of a fixed peak (analyzed peak) at various separations from a moveable "interference" peak of identical shape and size. The direct integration method simply used the area determined for a region, extending one FWHM on either side of the centroid of the analyzed peak. The (unfiltered) least-squares fit and the filter-fit curves were made with a single reference (for the analyzed peak). At zero separation, the analyzed and interference peaks overlap exactly and all three methods give twice the expected amplitude. As the separation increases to infinity, all three methods converge to unity. The filter-fit method approaches the "no-interference" condition the most slowly since convolution with the filter function broadens the effective width of a peak shape. Because of this extended response, it is important that either: (1) all peaks present in the spectrum be fitted, or (2) the fitting region be carefully selected so that nonreferenced peaks are excluded. This broadening effect increases with filter width, which is another practical reason for not using very large values for LW.

An interesting characteristic of the filter-fit response is that it exhibits regions of both positive and negative correlation. Thus, the presence of

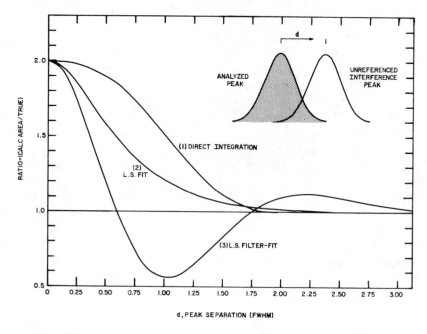

Figure 21.5 Response to an unreferenced peak for three analysis methods

an unreferenced peak might either increase or decrease the intensity determined for a nearby analyzed peak; there are even two nodes of orthogonality where an interfering peak does not affect the fitted amplitude of another peak at all. A large Chi-squared value, of course, will always be associated with the presence of such unreferenced peaks.

Sensitivity to Peak Shift

Figure 21.5 also indicates that the filter-fit method is very sensitive to peak shift since the curve drops very rapidly as a function of peak separation from "expected" position. This places stringent requirements on the stability of the detection electronics. With periodic spectrometer recalibration, users of the method have been able to operate from a single set of cassette-stored references for indefinite periods of time.[13] Instead of recalibrating the spectrometer, it is also possible to recalibrate an acquired spectrum. A gain and zero shifting program utilizing a simple quadratic interpolation technique has recently been developed for use in conjunction with the filter-fit program. A separate program determines the necessary gain and zero shift corrections by means of a two peak calibration. The results of early tests using this recalibration procedure have been very encouraging.

Summary of Some Published Results

The ML program incorporating the filter-fit method has been in use for over three years. During this time it has been employed with a variety of samples and excitation modes by users working in diverse applications of X-ray elemental analysis. Several workers involved in electron beam microanalysis have published results obtained with this program. Since the bremsstrahlung continuum component in such spectra is typically quite large and may be strongly modulated by matrix absorption, these analyses provide a particularly convincing demonstration of the capabilities of the method.

Beaman and Solosky have reported on the results of 79 quantitative elemental analyses;[14] the accuracy of the results was improved by the use of the filter-fit method. Corlett and McDonald have reported on the analysis of 13 minerals including difficult overlap situations such as PbS and HgS.[13] Results obtained using the ML program were found to compare favorably with those obtained by wet-chemical analysis.

Sargent, Meny and Champigny performed a comparison of quantitative analyses obtained with a scanning electron microscope (SEM) equipped with a silicon detector versus the results obtained on a microprobe equipped with crystal spectrometers.[15] The ML program was used for the unfolding of those spectra that contained peak overlaps. The accuracy of the SEM results was found to be comparable to those obtained with the microprobe.

Shuman and Somlyo applied the ML routine to the quantitative analysis of biological specimens in a transmission electron microscope.[16] They reported excellent results for the determination of potassium (a difficult problem due to calcium overlap and large bremsstrahlung continuum).

CONCLUSION

It has been shown that, by the application of a simple digital filter to both the unknown and the reference spectra, the conventional linear least-squares method can be made insensitive to smooth continuum. The technique has been successfully applied to the routine analysis of multi-element X-ray spectra acquired under a wide variety of excitation and sample conditions. Some of the significant advantages of this procedure are:

1. The need for explicit continuum models is eliminated.

2. Reference spectra are only required to be peak-shape standards and thus their preparation is greatly simplified.

3. The method adds a relatively small computational overhead to the least-squares procedure. It is suitable for implementation in a small on-line computer.

REFERENCES

1. Trombka, J. I. In *Applications of Computers to Nuclear and Radiochemistry*, NAS-NS 3107 (Washington, D.C.: Office of Technical Services, Department of Commerce, 1962), p. 183.

2. Salmon, L. In *Applications of Computers to Nuclear and Radiochemistry*, NAS-NS 3107 (Washington, D.C.: Office of Technical Services, Department of Commerce, 1962), p. 165.

3. Trombka, J. I. and R. L. Schmadebeck. *A Numerical Least-Square Method for Resolving Complex Pulse-Height Spectra*, NASA SP-3044 (Washington, D.C.: Office of Technology Utilization, National Aeronautics and Space Administration, 1968).

4. Gehrke, R. J. and R. C. Davies. "Spectrum Fitting Technique for Energy Dispersive X-Ray Analysis of Oxides and Silicates with Electron Microbeam Excitation," *Anal. Chem.* 47(9):1537 (1975).

5. Fiori, C. E., R. L. Myklebust, K. F. J. Heinrich and H. Yakowitz. "Prediction of Continuum Intensity in Energy-Dispersive X-Ray Microanalysis," *Anal. Chem.* 48(1):172 (1976).

6. Reed, S. J. B. and N. G. Ware. "Quantitative Electron Microprobe Analysis Using a Lithium Drifted Silicon Detector," *X-Ray Spectrom.* 2(2):69 (1973).

7. Smith, D. G. W., C. M. Gold and D. A. Tomlinson. "The Atomic Number Dependence of the X-Ray Continuum Intensity and the Practical Calculation of Background in Energy Dispersive Electron Microprobe Analysis," *X-Ray Spectrom.* 4(3):149 (1975).

8. Mariscotti, M. A. "A Method for Automatic Identification of Peaks in the Presence of Background and its Application to Spectrum Analysis," *Nucl. Instr. Meth.* 50:309 (1967).

9. Black, W. W. "Application of Correlation Techniques to Isolate Structure in Experimental Data," *Nucl. Instr. Meth.* 71:317 (1969).

10. Connelly, A. L. and W. W. Black. "Automatic Location and Area Determination of Photopeaks," *Nucl. Instr. Meth.* 82:141 (1970).

11. Robertson, A., W. V. Prestwich and T. J. Kennett. "An Automatic Peak-Extraction Technique," *Nucl. Instr. Meth.* 100:317 (1972).

12. Schamber, F. H. "A New Technique for Deconvolution of Complex X-Ray Energy Spectra," Proceedings of the Eighth National Conference on Electron Probe Analysis, New Orleans, Paper 85 (1973).

13. Corlett, M. I. and M. McDonald. "Quantitative Analysis of Sulphides and Sulfosalts Using an Energy Dispersive Spectrometer," Proceedings of the Ninth Annual Conference of the Microbeam Analysis Society, Ottawa, Canada, Paper 23, (1974).

14. Beaman, D. R. and L. F. Solosky. "Advances in Quantitative Energy Dispersive Spectrometry," Proceedings of the Ninth Annual Conference of the Microbeam Analysis Society, Ottawa, Canada, Paper 26 (1974).

15. Servant, J. M., L. Meny and M. Champigny. "Energy Dispersion Quantitative X-Ray Microanalysis on a Scanning Electron Microscope," *X-Ray Spectrom.* 4(3):99 (1975).

16. Shuman, H. and A. P. Somlyo. "Quantitative Electron Probe Analysis of Ultra-Thin Biological Specimens," Proceedings of the Tenth Annual Conference of the Microbeam Analysis Society, Las Vegas, Paper 41 (1975).

22

TRACE–A LEAST SQUARES FITTING PROGRAM
FOR PIXE SPECTRA

R. D. Willis, A. B. Baskin and R. L. Walter

Duke University and
Triangle Universities Nuclear Laboratory
Duke Station, North Carolina

INTRODUCTION

With increasing interest in rapid, multielement X-ray analysis techniques, attention has been directed toward the need for fast and reliable X-ray spectrum analysis programs. The level of sophistication in the most recent computer codes can be appreciated in papers by Kaufmann et al.[1] and Arinc et al.[2] representing sizeable programming efforts.

The purpose of this chapter is to describe the peak-fitting program TRACE developed for off-line analyses in conjunction with the particle-induced X-ray emission (PIXE) analysis system at Duke University. The primary emphasis of the PIXE studies at Duke has been multielement analyses of biological and environmental samples. Program TRACE was designed to permit rapid yet reliable analyses of spectra from a wide variety of sample types and compositions. The code, with the exception of some minor additions, was developed in the modest time of six weeks and requires approximately 14K of 24 bit memory words. The computation time for calculating fits to a spectrum of 16-20 elements is approximately one minute. The operation of TRACE in its present semiautomatic form requires an additional 30-60 sec per spectrum for operator interaction with the code. Future modifications, however, aimed at making TRACE a fully automatic code, will eliminate this phase of manual operation.

THE FITTING PROCEDURE

The characteristic features of a PIXE spectrum are indicated in Figure 22.1, which displays the X-ray spectrum from a PIXE irradiation of a thick-section of crab muscle. The plot shows the number of X-ray

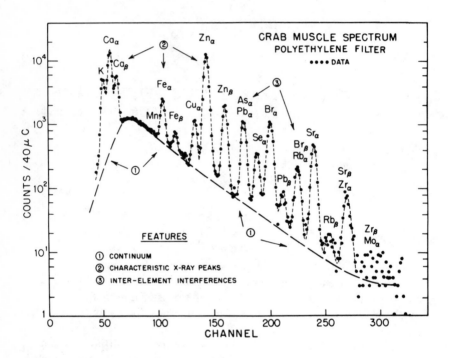

Figure 22.1 PIXE spectrum of crab muscle taken with a polyethylene filter. Characteristic features of PIXE spectra are noted.

counts observed as a function of energy or channel number for an accumulated charge of 40 μC. A polyethylene filter has been inserted between the Si(Li) detector and the scattering chamber to reduce the intensity of low-energy X-rays. The notable features of Figure 22.1 include a smoothly varying continuum (1), underlying characteristic X-ray peaks (2), and inter-element interferences (3), which are not resolved by the detection system. Not shown in the figure are detector escape peaks, direct pile-up peaks and the presence of low-energy tailing. These effects are associated with the presence of unusually large concentrations of one or more elements in the sample. Such concentrations are sufficiently rare in most of the biological samples analyzed at this laboratory that no corrections for these effects are presently included in TRACE. In those few exceptional cases in which these effects are not negligible, the corrections can be readily treated by extra manipulation, or, in the case of escape peaks and pile-up peaks, extra lines at the appropriate energies can be included in the fitting routine to generate fits to these peaks.

The Fitting Parameters

The fitting routine in TRACE seeks to minimize chi-square by the simultaneous adjustment of up to 13 fitting parameters. These parameters are stored in an array A and include the peak amplitudes of the Gaussian fits to the reference X-ray lines. The operator may choose to include any of 9 additional parameters within the constraint that A have a maximum of 13 elements. The function of these 9 parameters is discussed in more detail below, but briefly, these terms parameterize the calibration of the energy scale and the detector resolution (or peak width) and a five-term function describing the continuum shape.

Treatment of the Continuum

The current restriction on the size of the array A requires that most PIXE spectra be segmented into typically three or four regions each of which is independently fitted. By means of a light pen interactive scope display, the operator selects from the spectrum a set of data points that describe the continuum shape in the region being fitted. For routine cases, this selection process can be automatically handled by the computer eliminating this phase of operator interaction. The set of points are least-squares fitted to the five-term function BKGD(E,A):

$$BKGD(E,A) = EXP[A(0) + A(1) \cdot E + A(2) \cdot E^2 + A(3) \cdot E^3 + A(4) \cdot E^4]$$

Here, E is the energy (or equivalently, channel location) at which the background shape is evaluated. Any combination of the five background coefficients may be held fixed during the iterative fitting routine.

In most applications of the program, all five coefficients are fixed so that the background shape initially determined by the operator is not changed during the fitting process. By keeping the background fixed, the size of the array A (and therefore the computation time) is reduced, or the background coefficients in A may be replaced with five additional peak amplitudes allowing more elements in the spectrum to be fitted simultaneously. Figures 22.2 and 22.3 show the resulting TRACE fit to the crab muscle spectrum and to a spectrum of ashed bovine liver (NBS SRM 1577). The fitting regions indicated in Figure 22.2 are typical

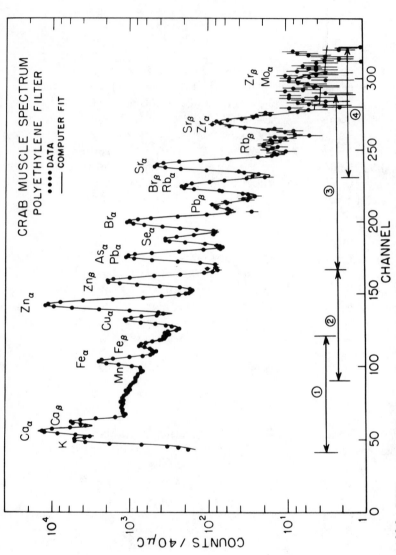

Figure 22.2   PIXE spectrum of crab muscle displaying computer fit to the data and indicating the separate fitting regions. Statistical uncertainties in the data are represented by vertical error bars.

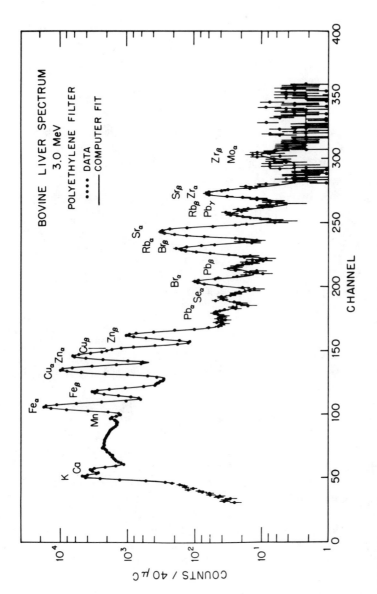

Figure 22.3 PIXE spectrum of ashed bovine liver (NBS SRM 1577) displaying computer fit to the data. Statistical uncertainties in the data are represented by vertical error bars.

segments that we select. They are chosen in this way primarily for convenience in fitting the background shapes.

Peak Fitting

The channel location X_0 in the pulse-height spectrum and the width Γ of a peak of energy E are determined by the linear relations:

$$X_0 = A(5) + A(6) \cdot E \qquad \text{and} \qquad \Gamma = A(7) + A(8) \cdot E.$$

The four coefficients A(5)-A(8) are normally included in the array A. Use of a fixed background permits up to nine elements (or approximately 18-25 individual lines) to be fitted simultaneously in a given segment of the spectrum.

After subtraction of the fitted background, the remaining peak spectrum is fitted with a set of Gaussian functions

$$PEAK(X,E,A)$$

where

$$PEAK(X,E,A) = RATIO(E) \cdot A(E) \cdot EXP \left[-\frac{1}{2} \left(\frac{X-X_0}{\Gamma} \right)^2 \right].$$

Here, X is the channel at which the function is evaluated, X_0 is the channel location of an X-ray line of energy E and width Γ, and $RATIO(E) \cdot A(E)$ is the amplitude of the X-ray peak where A(E) represents the amplitude of the reference peak and $RATIO(E)$ is the intensity ratio (determined from standards) for the given line. The present version of TRACE is supplied with energies and intensity ratios for 27 elements representing 58 X-ray lines. Additional lines are easily added if necessary In fitting the X-ray contributions from lead and mercury for example, TRACE normally uses only three Gaussians, one to fit each of the L_α, L_β, and L_γ peaks. For most biological samples, these are the only lines of lead and mercury to appear as significant peaks in the spectrum. In special cases, as in aerosol samples where unusually high concentrations of lead occur, the present capacity of TRACE can be expanded to include the additional spectral lines. Overlapping interferences as shown in Figure 22.1 are unfolded during the peak fitting routine using the known intensity ratios.

Minimization of Chi-Square

The computer-generated function FIT(X,E,A) is a superposition of the background function BKGD(X,E,A) and the collection of peak functions PEAK(X,E,A) for the region being fit:

$$FIT(X,E,A) = BKGD(X,E,A) + \Sigma \ PEAK(X,E,A).$$

The minimization of chi-square is based on the algorithm of Marquardt in which a least-squares fit is carried out after linearization of the fitting function.[3]

NUMERICAL RESULTS

Program TRACE is supplied with three experimentally determined efficiency tables (one table for each of the three absorbing filters used by this laboratory in collecting data), which convert peak areas of the Gaussian fits into absolute nanogram values for the corresponding elements. A second program is available to calculate correction factors for X-ray absorption and proton energy loss in thick sample irradiations for various matrix compositions. The minimum detectable limit for each element is calculated as 3σ where σ is the square root of the background counts beneath the element's reference peak. Finally, the reduced chi-square is calculated for each fit.

Figures 22.2 and 22.3 display the excellent agreement between the calculated TRACE fit and the original spectra. Vertical error bars through the data points represent the statistical uncertainties assigned to the data. The resulting abundances obtained from the fit in Figure 22.3 are tabulated in Table 22.1 and compared to the NBS certified values. The agreement is excellent with the exception of calcium and lead. The high lead value observed with PIXE is due in part to contamination of the acid used in the ashing procedure, and the low calcium value may indicate a need to readjust our calcium efficiency when using the polyethylene filter.

DISCUSSION AND CONCLUSION

The code TRACE is capable of calculating fits to 16-20 elements in approximately one minute. An additional amount of time, approximately one minute, is currently required in the manual selection of data points used to generate the background shapes in the different fitting regions. As mentioned earlier, for routine analysis this step in the fitting process can be handled by the computer in a negligible amount of time.

Table 22.1 Comparison of Duke PIXE Analysis of Bovine Liver with NBS Certified Values on ppm (Dry Weight) Basis

Element	Sample[a]				Sum[d]	Average Values	
	A	B	C	D		PIXE[b]	NBS[c]
K	0.90%	0.93%	0.89%	1.01%	0.90%	0.93(0.05)%	0.97(0.06)%
Ca	—	—	—	—	100	100	123
Cr	1.3	2.6	2.7	0	1.6	2.4(0.7)	
Mn	10.9	10.5	8.5	8.0	8.6	9.5(1.4)	10.3(1.0)
Fe	270	282	287	310	272	287(17)	270(20)
Ni	0.8	0.8	0.1	1.2	0.4	0.7(0.5)	
Cu	186	196	193	207	183	196(8)	193(10)
Zn	140	142	146	152	138	145(5)	130(10)
Ga	1.0	1.0	0.3	1.9	0.8	1.1(0.7)	
As	<0.6	<0.7	<0.7	1.0	<0.3	<0.7	0.055
Se	1.2	1.3	1.2	1.1	1.2	1.2(0.1)	1.1(0.1)
Br	5.6	7.1	5.1	6.6	5.9	6.1(0.8)	
Rb	20	18	17	17	18	18(1.0)	18.3(1.0)
Mo	2.3	5.2	1.9	4.9	3.4	3.5(1.5)	3.2
Pb	4.5	2.3	4.5	4.6	3.5	3.9(1.0)	0.34(0.08)

[a]Samples are acid-ashed and represent 0.13-0.15 g of bovine liver.

[b]PIXE values are based on internal strontium standard.

[c]NBS values that are listed with uncertainties (given in parenthesis) have been certified.

[d]Sum represents values obtained by analysis of the composite spectrum obtained by adding the spectra from samples A, B, C and D.

In conclusion, the authors wish to emphasize that for a modest programming effort, one can obtain fast and reliable analyses of X-ray spectra.

REFERENCES

1. Kaufmann, H. C., K. R. Akselsson and W. J. Courtney. *Advances in X-Ray Analysis*, Vol. 19 (Dubuque, Iowa: Kendall/Hunt Publishing Co., 1976), p. 355.
2. Arinc, F., L. Wielopolski and R. P. Gardner. "The Linear Least-Squares Analysis of X-Ray Fluorescence Spectra of Aerosol Samples Using Pure Element Library Standards and Photon Excitation," Chapter 20 this volume.
3. Bevington, P. R. *Data Reduction and Error Analysis for the Physical Sciences* (New York: McGraw-Hill, 1969), pp. 235-245.

23

DETERMINATION OF THE BACKGROUND IN PROTON-EXCITED X-RAY SPECTRA IN THE ABSENCE OF BLANKS

R. A. Semmler

IIT Research Institute
Chicago, Illinois

INTRODUCTION

While most reports on spectrum analysis have emphasized the problem of finding and fitting the X-ray peaks, the background underneath the peaks must also be determined for accurate analysis. In routine analytical applications, the problem is often solved by analysis and subtraction of a blank from the data. However, in the analysis of thick one-of-a-kind samples, X-ray spectra from proton excitation are produced with no possibility of comparison with a blank. While proton-excited spectra are somewhat unique, many of the comments apply to spectrum analysis in general. The only other known survey is given by Stratham.[1]

FACTORS CONTRIBUTING TO THE BACKGROUND

In addition to the characteristic X-ray lines, the X-ray spectra produced by beams of heavy charged particles may contain components from a variety of sources such as: (1) bremsstrahlung from the incident particle, (2) bremsstrahlung from secondary electrons, (3) gamma radiation from nuclear reactions induced in the target, and (4) bremsstrahlung from free electrons collected by the charged sample. A complete analysis of these components and others has been given by Folkmann.[2,3] Normally, the dominant component is bremsstrahlung from secondary electrons, but other components are not negligible and will vary in importance depending on target thickness, conductivity, and atomic number.

BACKGROUND ESTIMATION TECHNIQUES

A variety of techniques has been used to estimate the background underneath an X-ray line. While the problem is trivial in the case of a single line on a flat background, it becomes difficult for the case of multiple overlapping lines on a nonlinear background. A typical proton-excited X-ray spectrum is shown in Figure 23.1. Some of the techniques that have been reported for determining the background are reviewed briefly below. Generally, the emphasis here is on techniques that require no operator intervention and no prior knowledge of the elements present (such as required for library fitting techniques).

Measurement of a Blank

In the analysis of filter samples, measurement of a clean blank is a common method of determining the background level. For this type of sample, the technique is simple, direct, and generally reliable.[4]

Peak Clipping

Peak clipping techniques have been suggested by Ralston and Wilcox.[5] The method involves a comparison of the intensity of each point in the spectrum with the average intensity in the local region. If the actual value deviates significantly from the local average, this is interpreted as a bump or peak, and the actual value is replaced by the average. The effect is to clip the tops off any peaks present, much as if one were sanding wood. The process is repeated until all significant deviations are removed. The technique works best if the peaks are sharp and pronounced when compared to variations in the background. The method works better with typical Ge(Li) gamma ray spectra than with Si(Li) X-ray spectra, but the technique is applicable to both.

Library Routines

Generally we have avoided techniques that require subtraction of, or comparison with, a library of pure element spectra because of the need for *a priori* knowledge of the probable composition as well as a need for highly reproducible gain and zero settings. However, several variations of this technique should be noted: (1) "iterative stripping" which subtracts out only a portion of each peak on each of several passes through the

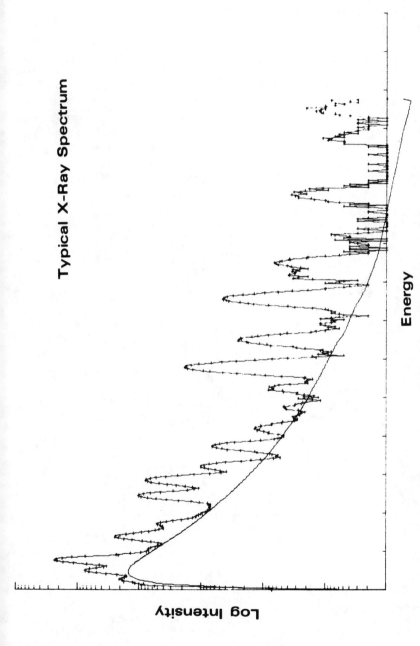

Figure 23.1   A typical proton-excited X-ray spectrum collected with a Si(Li) detector.

spectrum,[1]   (2) library fitting routines that fit the first derivative of the library standard to the first derivative of the unknown,[6] and (3) library fitting routines that fit the second derivative of the standard to the second derivative of the unknown.[7]

Arbitrary Polynomial Functions

Arbitrary analytical functions such as orthogonal polynomials have been used for many years to represent an unknown background component in least squares fitting routines for gamma ray spectra.[8]   The computer required, however, is generally large because of the number of parameters required.   Polynomials of ninth degree have been required to fit even small portions of a spectrum.

Figures 23.2 and 23.3 are an example of background determination using an arbitrary polynomial.   The background has a shape similar to one determined from a clean blank except that there is a tendency for

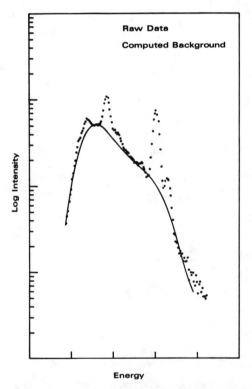

Figure 23.2 Background polynomial function (6th degree) determined by nonlinear least squares program simultaneously with parameters for seven Gaussian peaks.

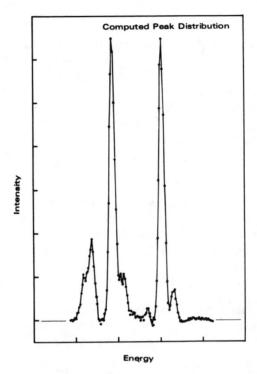

Figure 23.3   Peak structure from spectrum in Figure 23.2 after polynomial background subtraction.

the background to bulge in regions beneath a cluster of peaks. Even for this limited portion of a spectrum, a sixth degree polynomial was required by the least-squares fitting routine in addition to the seven peaks, *i.e.,* a total of 28 parameters. Also, logarithmic or other compressed scales are essential to minimize the polynomial degree required for a good fit.

Houston has reported on a modified polynomial technique that relies on multiple differentiation followed by multiple integration to eliminate the slowly varying background without destroying the peak structure.[9] This is similar to the approach of Schamber[7] but eliminates the need for a library. The primary difficulty in applying this approach to X-ray applications would appear to be the instability of higher order numerical derivatives.

Special Functions

Specific simple analytic functions with only a few parameters such as a straight line, parabola,[10] or Fermi-like distribution[11] have been used within limited regions. The entire background contribution in X-ray spectra has also been represented by specific analytic functions, but this generally requires the use of six or more parameters.[12] In many cases, we have found the following three parameter descriptions to adequately represent the normal background shape:

$$I = AE^n e^{-\lambda E} \tag{23.1}$$

where A, n and λ are arbitrary parameters that primarily determine the amplitude, low energy roll-off, and high energy tail respectively. The primary argument against any of these forms is that it forces the background to fit a preconceived and possibly erroneous shape.

Empirical Techniques

Empirical techniques for estimating the background in a spectrum have also been developed. These usually involve modifications to the line initially determined by connecting valleys in the spectrum.[13]

Filtered Raw Data

While work in the transform domain is more common, filters have also been applied directly to raw data. For example, application of the filter reported by Volkov[14] will erode the peak structure somewhat as do the peak clipping techniques.

Filtered Fourier Transforms

Fourier transform techniques have been applied to spectrum analysis for smoothing[13] and peak enhancement.[15] With such techniques, the primary step is to represent the raw spectral data by a summation of sine and cosine waves (see Figure 23.4). The new representation is an advantage only if one can perform spectral modifications either faster or easier.

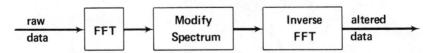

Figure 23.4 Block diagram representing the primary steps in spectrum alteration using Fourier transforms.

For example, one can compute weighted averages for a spectrum using the raw data values within a sliding window. Somewhat surprisingly, the same result can be achieved in the same time or faster by converting the data to a mixture of oscillating waves, modifying the intensity of individual waves, and then reconstructing the spectrum. Also, the viewpoint provided by the new representation may suggest new methods of spectral analysis. For example, if the new picture consisted of three different groups of frequencies representing background, data peaks, and random noise, the isolation of each important component would be easy. Unfortunately, some of the same frequencies are required in all three groups, so that one must compromise when attempting to extract or filter out a specific component in the spectrum by isolating specific frequencies. The contribution of each component is shown in Figure 23.5.

The concept of filtering or modifying the frequency components is basic to most applications of transform theory. Extensive literature on the definition and implementation of filters to isolate specific signals is available but provides no more than a starting point. One common approach to extraction of a signal with known characteristics is the so-called Wiener filter, which minimizes the mean square error between the actual and desired filter output. The Wiener filter is defined by the transfer function $W = P_S/(P_S + P_N)$ where P_S and P_N are the power spectra for the input signal and input noise components respectively. The transform Y of the filter output is obtained from $Y = WX$ where X is the transform of the observed (signal plus noise) input. The inverse transform of Y gives the filter output.

EXAMPLES

As an example of some of these techniques, one can define a Wiener filter that ideally would accept raw data and transmit only a background function. If the background is estimated as in Figure 23.1 (actually done with a peak clipping routine) and used as a starting point to define a background-pass filter, the actual output of this filter will be the curve shown in Figure 23.6. Substantial oscillations are present compared to the estimate.

One possibility for refinement would be to use this curve as the comparison curve in a peak clipping routine. The family of the curves shown in Figure 23.6 illustrates the results of repeated comparison of the filter output in a peak clipping program. The final curve generally tracks along the expected background line except at low energies where the rolloff appears excessive. The rolloff at low energy is caused by the inability to follow the initial sharp rise because the necessary high frequency

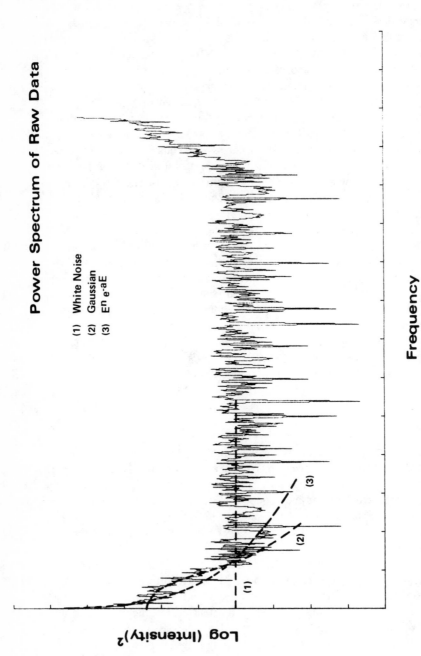

Figure 23.5 Power spectrum of the data shown in Figure 23.1. The dashed lines indicate schematically the three major sources of the observed structure: (1) random noise, (2) Gaussian peaks, and (3) background. (The left and right halves are symmetric.)

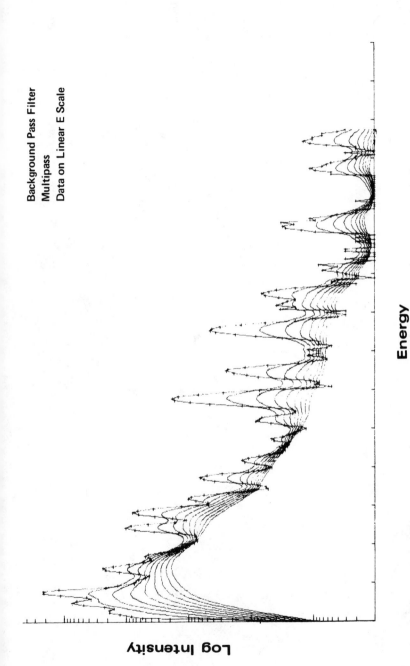

Figure 23.6 Top curve (with crosses) is raw data from Figure 23.1. Next lower curve is output of Wiener filter defined using estimated background in Figure 23.1. Subsequent curves are result of combining the filter output with a peak clipping program.

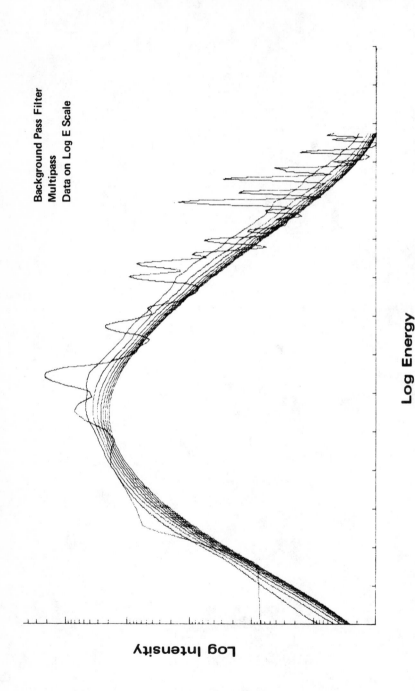

Figure 23.7 Top curve is same raw data as in Figure 23.1 but on a logarithmic energy scale. Next lower curve is output of Wiener filter defined using estimated background in Figure 23.1. Subsequent curves are result of combining the filter output with a peak clipping program.

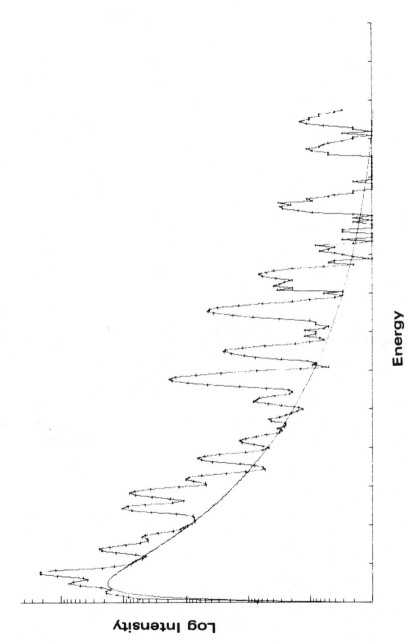

Figure 23.8 The final background curve from Figure 23.7 after retransformation back to a linear energy scale.

components are not passed by the filter. A possible cure for this characteristic is to precondition the data by transforming to a logarithmic energy scale in order to eliminate the abrupt leading edge. After redefining the filter using this preconditioned data, the initial filter output and the family of curves generated by a peak clipping routine are shown in Figure 23.7. Most of the oscillations have been eliminated by avoiding the sharp leading edge. The final curve transformed back to a linear energy scale and compared with the raw data is shown in Figure 23.8.

This particular approach is not unique and an almost endless combination of filters and preconditioning programs can be defined, compared, and optimized. For example, an obviously beneficial transformation would compress the energy axis only enough to result in peaks of constant width. The frequency components representing the Gaussian peaks would then represent all peaks equally well rather than the average peak. In view of these possibilities, filtering techniques still need much development. However, the technique is inherently flexible and powerful and will undoubtedly receive further attention.

SUMMARY

Judging from the experience with gamma ray spectroscopy, X-ray spectrum analysis techniques will continue to be reported. Often, results will largely duplicate previous work but with some special feature added. The particular techniques that appear to be most useful at this time for background determination are the peak clipping technique because of its simplicity, and the filtered transform technique because of its flexibility. The selection of the best filter or definition of an algorithm to pick an appropriate filter in an arbitrary case are problems only partially answered at present.

REFERENCES

1. Stratham, P. J. "A Comparative Study of Techniques for Quantitative Analysis of the X-Ray Spectra Obtained with a Si(Li) Detector," *X-Ray Spectrometry* 5:16-28 (1976).
2. Folkmann, F., C. Gaarde, T. Huus and K. Kemp. "Proton-Induced X-Ray Emission as a Tool for Trace Element Analysis," *Nucl. Instr. Methods* 116:487-499 (1974).
3. Folkmann, F., J. Borggreen and A. Kjeldgaard. "Sensitivity in Trace-Element Analysis by p, α and ^{16}O Induced X-Rays," *Nucl. Instr. Methods* 119:117-123 (1974).
4. Harrison, J. F., and R. A. Eldred. "Automatic Data Acquisition and Reduction for Elemental Analysis of Aerosol Samples," in *Advances in X-Ray Analysis*, Vol. 17, C. L. Grant, Charles S.

Barrett, John B. Newkirk and Clayton O. Ruud, Eds. (New York: Plenum Press, 1974), pp. 560-570.

5. Ralston, H. R. and G. E. Wilcox. "A Computer Method of Peak Area Determinations from Ge(Li) Gamma Spectra," in *Modern Trends in Activation Analysis, National Bureau of Standards Special Publication 312*, Vol. II, James R. Devoe, Ed. (Washington, D.C.: U.S. Government Printing Office, 1969).

6. Brouwer, G. and J. A. J. Jansen. "Deconvolution Method for Identification of Peaks in Digitized Spectra," *Anal. Chem.* 45:2239-2247 (November 1973).

7. Schamber, F. "A Modification of the Linear Least Squares Fitting Method Which Provides Continuum Supression," Chapter 21 this volume.

8. Stevens, M. M. and J. A. Harvey. "Analysis of Neutron-Capture Gamma-Ray Spectra," in *Slow-Neutron-Capture Gamma Ray Spectroscopy, Argonne National Laboratory Report No. ANL-7282*, F. E. Thyow, Ed. (Springfield, Va.: Clearinghouse for Federal Scientific and Technical Information, 1968).

9. Houston, J. E. "Dynamic Background Subtraction and the Retrieval of Threshold Signals," *Rev. Sci. Instr.* 45:897-903 (1974).

10. Routti, J. T. and S. G. Prussin. "Photopeak Method for the Computer Analysis of Gamma-Ray Spectra from Semiconductor Detectors," *Nucl. Instr. Methods* 72:125-142 (1969).

11. Teoh, W. "Cutipie—A Computer Program to Analyze Gamma-Ray Spectra," *Nucl. Instr. Methods* 109:509-513 (1973).

12. Kaufman, H. C. and R. Akselsson. "Non-Linear Least Squares Analysis of Proton-Induced X-Ray Emission," in *Advances in X-Ray Analysis*, Vol. 18, William L. Pickles, Charles S. Barrett, John B. Newkirk and Clayton O. Ruud, Eds. (New York: Plenum Press, 1975), pp. 353-361.

13. Inouye, T. "Application of Fourier Transforms to the Analysis of Spectral Data," *Nucl. Instr. Methods* 67, 125-132 (1969).

14. Volkov, N. G. "Subtraction of the Background and Automatic Peak Identification in Gamma-Ray Spectra Obtained from a Ge(Li) Detector," *Nucl. Instr. Methods* 113(4):483-488 (1973).

15. Horlick, G. "Resolution Enhancement of Line Emission Spectra by Deconvolution," *Appl. Spectroscopy* 26:395-399 (1972).

24

CORRECTION FOR LINE INTERFERENCES
IN WAVELENGTH-DISPERSIVE X-RAY ANALYSIS

J. V. Gilfrich, L. S. Birks and J. W. Criss

Naval Research Laboratory
Washington, D.C.

INTRODUCTION

In wavelength-dispersive X-ray analysis, interferences between lines from different elements are not a severe problem. Conventional crystal spectrometers are capable of resolution sufficient to completely separate most lines of interest. There do exist, however, certain cases where line interference can occur and where corrections must be made to obtain an accurate analytical result.

During the calibration procedure, it is a straightforward step to establish a set of interference coefficients that can be used in an iterative fashion to obtain the corrected intensity for each analytical line. Each interference coefficient is simply the intensity of the interfering element measured at the affected line position, divided by the intensity measured at the peak position for the interfering element itself.

DETERMINATION OF INTERFERENCE COEFFICIENTS

The best way to describe the simplicity of this correction is by the use of a specific case. Figure 24.1 shows the case chosen, the Cr K_β-Mn K_α doublet, as it might appear if recorded using a crystal spectrometer having about 20 eV resolution. The dashed lines in the illustration indicate that each of the lines contributes some intensity to the other. It must be borne in mind that wavelength-dispersive measurements are

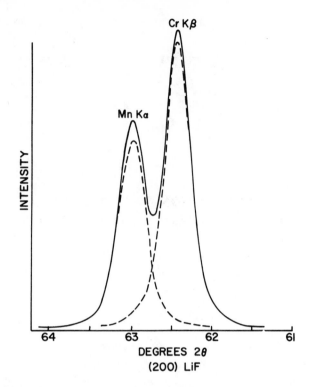

Figure 24.1 Mn K$_\alpha$–Cr K$_\beta$ doublet recorded with a crystal spectrometer having 20 eV resolution.

conventionally made by positioning the spectrometer at the peak of the line and counting for a fixed period. An analyst would not ordinarily scan the line to produce the peaks as shown in this figure.

Through the use of single-element standards (or noninterfering multi-element standards) it is possible to determine the coefficients that define the contribution of each element to the peak position of the other element(s), as a part of the sensitivity calibration. Figure 24.2 represents the two single element standards, chromium and manganese, showing the intensity from chromium at the Mn K$_\alpha$ position and from manganese at the Cr K$_\beta$ position. In a sequential crystal spectrometer it is a simple matter to measure these interfering intensities. In Figure 24.2, simply scaling the intensities shows that the effect of chromium on the Mn K$_\alpha$ peak position is 5% of the Cr K$_\beta$ intensity and that the effect of manganese on the Cr K$_\beta$ position is 6% of the Mn K$_\alpha$ intensity.

There may be circumstances where it is not possible to measure the Cr K$_\beta$ intensity; for example, in a simultaneous multicrystal instrument,

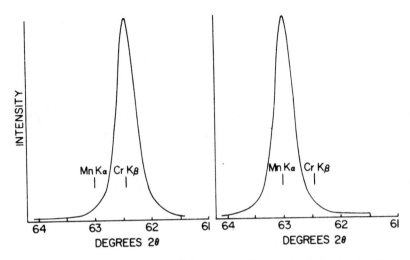

Figure 24.2 Mn K_α and Cr K_β from single element standards showing the measurement of interference coefficients.

the measurement of chromium undoubtedly would use the Cr K_α line. Since the "thin sample" being treated here always has the same K_α/K_β ratio (unlike the bulk sample case, where the ratio may be dependent on the other elements present), the interference coefficient of chromium on manganese can be expressed as a fraction of the Cr K_α line. Assuming that other elements which might interfere with Cr K_α are absent, the correction becomes a simple matter of subtracting the product of the interference coefficient and the Cr K_α intensity from the Mn K_α intensity; iteration is not required, as will be illustrated below.

APPLICATION OF THE INTERFERENCE COEFFICIENTS

The correction procedure involves merely the solution of a set of simultaneous equations such as:

$$I_i^M = I_i^T + f_{ij} I_j^T \tag{24.1}$$

for each element i, where

i = the affected element
j = the interfering element
I_i^M = the measured intensity of element i
I_i^T and I_j^T = the true intensities of elements i and j, respectively
f_{ij} = the interference coefficient for the effect on element i by element j.

Since I_i^T is the desired quantity, the set of simultaneous equations are rearranged to the form:

$$I_i^T = I_i^M - f_{ij} I_j^T \qquad (24.2)$$

and solved iteratively for I_i^T, by substituting I_j^M for I_j^T on the right side of the equations in the first iteration, and substituting the current value of I_j^T on subsequent iterations. For the case described at the end of the previous section, where f_{MnCr} represents a fraction of the Cr K_α intensity, I_{Cr}^T is known and no iteration is necessary because we have assumed that there is no interference with the Cr K_α line.

The best way to describe the iterative procedure is by example. As was determined when discussing Figure 24.2:

$$f_{MnCr} = 0.05; \quad f_{CrMn} = 0.06$$

Let us now assume that

$$I_{Mn\ K_\alpha}^M = 400 \text{ c/s}; \quad I_{Cr\ K_\beta}^M = 600 \text{ c/s}$$

For the first iteration take $I_{Mn\ K_\alpha}^T = I_{Mn\ K_\alpha}^M$ and $I_{Cr\ K_\beta}^T = I_{Cr\ K_\beta}^M$ on the right side of equations like Equation 24.2:

1. $\quad I_{Mn\ K_\alpha}^T = I_{Mn\ K_\alpha}^M - f_{MnCr} I_{Cr\ K_\beta}^M$

    $= 400 - (0.05)(600) = 370 \text{ c/s}$

   $I_{Cr\ K_\beta}^T = I_{Cr\ K_\beta}^M - f_{CrMn} I_{Mn\ K_\alpha}^M$

    $= 600 - (0.06)(400) = 576 \text{ c/s}$

In each succeeding iteration use, on the right side of the equations, the "true" intensities determined in the previous iteration:

2. $\quad I_{Mn\ K_\alpha} = 400 - (0.05)(576) = 371.2 \text{ c/s}$

   $I_{Cr\ K_\beta} = 600 - (0.06)(370) = 577.8 \text{ c/s}$

3. $\quad I_{Mn\ K_\alpha} = 400 - (0.05)(577.8) = 371.1 \text{ c/s}$

   $I_{Cr\ K_\beta} = 600 - (0.06)(371.2) = 577.7 \text{ c/s}$

4. $I_{Mn\ K_\alpha}$ = 400 - (0.05) (577.7) = 371.1 c/s

   $I_{Cr\ K_\beta}$ = 600 - (0.06) (371.1) = 577.7 c/s

In this simple case it can be seen that the true values for $I_{Mn\ K_\alpha}$ and $I_{Cr\ K_\beta}$ were reached after three iterations, as evidenced by no change in the fourth. Convergence criteria could be established that might require even fewer, or in isolated cases, more iterations, or the process could be stopped after a fixed number of steps.

The case of only one line suffering interference can be illustrated using values consistent with the above, assuming that Cr K_α is five times as intense as Cr K_β; f_{MnCr} becomes 0.05/5 = 0.01, Cr K_α intensity becomes 2888 c/s (5 x 577.7), and $I_{Mn\ K_\alpha}$ = 400 - (0.01 x 2888) = 371.1 c/s the same value determined above.

As an example of the necessity for making these corrections, it is appropriate to cite the experience with the multicrystal simultaneous wavelength dispersive instrument in use in the EPA Environmental Sciences Research Laboratory at Research Triangle Park.[1] This instrument has 16 fixed channels plus a scanning spectrometer and has been programmed to measure 33 elements. It might be thought that setting up a 33 x 32 matrix to account for 1056 possible interferences would be desirable. But most elements do not suffer any interferences at all (above the 0.1% level) and so most of the coefficients in such a matrix would be zero. In the measurement of 33 elements on the EPA instrument 11 elements are affected by interferences, requiring the storage of 19 coefficients in the computer that controls the instrument and reduces the data, automatically applying the iterative correction. A few of these coefficients are listed in Table 24.1, to illustrate that one element can interfere with more than one other element and that one element can be affected by more than one other element, and to show some typical values.

Table 24.1 Interference Coefficients

Interfering Element	Affected Element	Interference Coefficient
Sb L_α	Ca K_α	0.0012
Sb L_α	Sn L_α	0.0020
Ba L_α	Ti K_α	0.0063
Pb M_α	S K_α	0.013
Br K_α	As K_β	0.074
W L_γ	As K_β	0.156
Hg L_β	As K_β	0.443

SUMMARY

In spite of the good resolution of a crystal spectrometer, there are some line interferences that must be corrected in order to obtain accurate X-ray intensities. Because the interferences are few, intensities can be corrected in a simple manner by the incorporation of an interference coefficient. These coefficients can be determined in a straight-forward way during the calibration of the spectrometer and applied in a routine manner during data reduction.

REFERENCE

1.   Wagman, J.   Environmental Sciences Research Laboratory, EPA, Research Triangle, Park, North Carolina, private communication.

25

DETECTION AND QUANTITATION
IN X-RAY FLUORESCENCE SPECTROMETRY*

Lloyd A. Currie

Analytical Chemistry Division
National Bureau of Standards
Washington, D.C.

INTRODUCTION

The two-fold objective of this chapter is to consider: (1) criteria for assessing the performance characteristics of the various X-ray fluorescence methods used for the analysis of environmental samples, and (2) means for unambiguously reporting the results of such measurements. The primary performance characteristics to be treated here will be limits for detection and quantitation. These characteristics are determined by the sensitivity (calibration) function, systematic error bounds, counting errors, and other sources of random error. For experimental results, the principal question is: how best to express those results that lie near or below the limit of detection. Conclusions and recommendations will be summarized in the final section of the text.

To focus on some of the issues, let us examine some recent experimental results. The upper section of Table 25.1 shows intercomparison results from nine experienced laboratories for the analysis of trace mercury levels in an NBS Standard Reference Material (SRM 1577-Bovine Liver).[1] The intercomparison exercise, which took place in connection with the International Decade of Ocean Exploration, included a number

---

Table 25.1 Reporting Practices (Intercomparison Results)

Hg (μg/g) in SRM 1577 (bovine liver)	
0.055 ± 0.014	0.016 ± 0.002
0.070 ± 0.015	0.012 ± 0.005
≤ 0.2	0.021
0.043 ± 0.024	0.09
	0.006 ± 0.0006

SRM 1633 (coal fly ash)	
As (μg/g, lab 11):	<100, <100, <100, <100, <100
S (%, lab 9):	<5.5, <6.7, <6.0, <5.0, <6.0

of different trace analysis methods. Examination of the data leads immediately to certain questions related to quoted experimental errors. These are: What are the meanings of the uncertainties? Do they represent standard deviations, standard errors, or some multiple thereof? Are they derived from Poisson statistics or from replication? What exactly was repeated and how many times? What is the significance of the upper limit? Does it represent a detection limit? How defined? Does it involve some multiple of the standard deviation? How are we to interpret results lacking uncertainty bounds?

Partial information was provided to help answer some of these questions.[1] Three of the nine laboratories furnished the number of degrees of freedom or the number of replicates; two stated that upper limits would be defined as 2σ and 6σ; and four stated their uncertainty recipes: standard error (one laboratory), standard deviation (two laboratories), and twice the standard deviation (one laboratory). Information on bounds for the systematic error was not evident for any of the reporting groups.

The second example, extracted from an NBS-EPA intercomparison exercise, consists of individual replicate results quoted by two of the laboratories that measured trace elements in NBS SRM 1633-coal fly ash.[2] These values are in the bottom half of Table 25.1. Obviously, neither lab had adequate detection power to measure the element in question. Most of the other participants were able to detect arsenic (certified concentration, 61 ± 6 micrograms/gram) and sulfur (approximate concentration, 0.3 wt/%). Besides wondering about the specific upper limit definitions used by the two laboratories, we note here two additional features. First, *different* definitions were surely used: laboratory #11 shows the same limit for all replications, while laboratory #9 shows a variable upper limit. Second, information was lost by reporting nothing more than a sequence of limits. The individual results and their standard deviations could have been combined in each case to yield a mean or a smaller upper limit. The power of averaging to reduce the standard error by a factor of $\sqrt{5}$ has been lost.

The purpose of these examples is merely to illustrate the continuing need for analysts to employ explicit, and preferably accepted, means for reporting experimental results and for assessing method capabilities. The problem becomes acute when the values measured are at or below the detection limit.

In the following sections, we shall first review basic concepts underlying detection and quantitation limits and data reporting in chemical analysis, and then apply these principles to X-ray fluorescence measurements. Initially, only Poisson counting statistics will be considered; then, additional sources of imprecision and systematic error will be treated. Following a comparative assessment of XRF measurement adequacy for a number of airborne particulate pollutants, equations will be presented that allow the objective evaluation of alternative methods and the unambiguous reporting of detection limits.

A number of sources contain valuable information on measurement needs and capabilities of the various XRF methods. Dzubay and Stevens[3] compared the detection limits for an energy dispersive system with typical trace element concentrations in airborne particles for most of the periodic table. An intercomparison of the performance of wavelength and energy dispersive XRF plus three other analytical techniques for synthetic and natural aerosol particle samples has been published by Camp et al.[4] Attainable accuracy as well as important sources of bias were revealed through this exercise. Information concerning performance characteristics and error sources for proton-induced X-ray emission has been discussed recently by Campbell et al.[5] Considerable insight into sources of systematic error and methods of correction in electron probe and X-ray analysis has been provided by Heinrich.[6,7] The comprehensive treatment of detection limits in XRF in the book by Müller[8] is noteworthy. Müller also provides much useful information on detectability in the form of summaries of relevant research papers.

BASIC CONCEPTS AND DEFINITIONS

Detection and Quantitation

Two concepts that are particularly valuable for expressing the measurement capabilities of an analytical method are the detection limit C_D and the quantitation limit C_Q. In order to express C_D and C_Q as useful numerical quantities it is necessary to have explicit algebraic definitions. Following the development in Reference 9 we shall define C_Q as the concentration for which the relative standard deviation (RSD) of the measured concentration is 10%. The detection limit C_D, on the other hand, is defined as the smallest concentration that a particular measurement process can reliably detect. Detection may be considered "reliable" when the probability of detecting concentration C_D (1-β) is sufficiently large, while the probability of spuriously detecting a blank (α) is sufficiently small. Utilizing nomenclature derived from Reference 10 we shall label these probabilities as detection power (1-β) and false positive risk (α); they will be assigned values of 95% and 5%, respectively.

Explicit definition of C_D requires us not only to assign numerical values to α and β, but also to specify rules according to which the decision "detected" (d) or "not detected" (nd) is to be made. As indicated in Figure 25.1 the decision may be made by comparing an experimental result with a decision limit C_C so constructed that the false positive risk is equal to α. That is, if the chemical element in question is, in fact, absent (C = 0), the probability that a concentration measurement would exceed C_C is only 5% (α). Similarly, measurement of a substance whose true concentration is C_D has a 95% chance (1-β) of exceeding C_C.

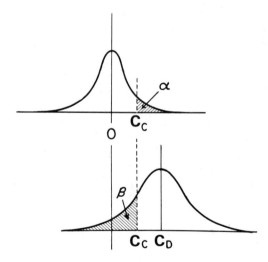

Figure 25.1 Illustration of the meaning of C_C and C_D. The upper curve shows a distribution of measured concentrations when the true concentration is zero. The lower curve illustrates a distribution of measured concentrations when the true concentration equals C_D. The decision "detected" is made when an experimental result exceeds C_C.

Table 25.2 will assist in further clarifying the relationships; the probabilities for correct $(1-\alpha, 1-\beta)$ and incorrect (α,β) decisions are shown at the four intersections in the table. Note that the decision d or nd characterizes an experimental result, but the detection limit C_D characterizes the analytical procedure. C_C is merely a numerical quantity with which the experimental result $(\widehat{C})$ is judged. If $\widehat{C} > C_C$, one should decide d, otherwise, nd. If the error distribution is symmetric and $\alpha = \beta$, $C_C \approx \frac{1}{2}C_D$. Some investigators prefer to report nd or give an upper limit only when $\widehat{C} < 0$ (rather than $\widehat{C} \leqslant C_C$). This is foolhardy, for it means that blanks will be "detected" 50% of the time.

Table 25.2   Truth Table[a]

Decision	Actual Concentration C	
	Blank (C = 0)	Detection Limit (C = C_D)
nd $(\widehat{C} \leqslant C_C)$	$1 - \alpha$	β
d $(\widehat{C} > C_C)$	α	$1 - \beta$

[a]Body of the table gives decision probabilities. C_C is characterized by by false positive risk α (5%); C_D is characterized by C_C (or α) and the false negative risk β (also 5%).

Computation of C_D and C_Q

The above theory (hypothesis testing) is valid so long as all errors are random. Numerical values may be assigned once the form of the random error distribution is assumed. Lest these assumptions seem too severe ever to be of use in practice, it is worth remembering that the validity of statistical confidence intervals rests upon *exactly* the same assumptions. Assuming normality (Gaussian distribution), taking $\alpha = 0.05 = \beta$ and ignoring small variations of standard deviation with signal level below the detection limit, we obtain:[9]

$$decision \quad C_C = z_{1-\alpha}\, \sigma_0 = 1.64\, \sigma_0 \qquad (25.1)$$

$$detection* \quad C_D \approx 2\, C_C = 3.29\, \sigma_0 \qquad (25.2)$$

$$quantitation \quad C_Q = 10\, f\, \sigma_0 \qquad (25.3)$$

*It is fortuitous, and potentially misleading, that both $\sqrt{2}\, z_{0.99} = \sqrt{2}$ (2.326) and $2z_{0.95} = 2$ (1.645) equal 3.29.

where $z_{1-\alpha}$ is the abscissa of the normal distribution corresponding to $\alpha = 0.05$ (Figure 25.1); and σ_0 is the standard deviation of the observed result ($\widehat{C}$) when the true concentration is zero (a blank). The factor f, which is generally close to unity, is given by equation 25.9 in the section on Poisson counting statistics. If σ_0 were derived from M measurements of the blank, it would be replaced by the estimated standard deviation in the above expressions, and z would be replaced by Student's-t.[10]

The crucial factor in Equations 25.1-25.3 is σ_0, which is related to the standard deviation of the blank σ_b by

$$\sigma_0 = \sigma_b \sqrt{\eta} \qquad (25.4)$$

where $\eta = 1$ if the background (or blank) is well characterized ($M \gg 1$), and $\eta = 2$ if paired observations, (at equal counting times) of sample and blank are made.[9] When least-squares spectrum fitting is employed, σ_0 may be determined directly from the variance-covariance error matrix for that method.[11] Allowance must be made for the propagation of random errors associated with measurements of the sample, interfering species, the blank, correction factors, and the calibration function. In the next section σ_0 will be interpreted in terms of counting statistics.

Reporting of Trace Analytical Data

Deficiencies in reporting clarity were evident from Table 25.1. The reporting of large (d) signals often suffers from incomplete information concerning the standard deviation multiplier, estimated bounds for bias and/or numbers of degrees of freedom. The reader is urged to examine the paper by Eisenhart on this topic.[12] For small (nd) signals the situation is worse.

When an experimental result falls into the nd category (i.e., $\widehat{C} \leqslant C_C$) four reporting options exist:

(1) nd
(2) $< (\widehat{C} + k \ \sigma)$
(3) $< k'\sigma$
(4) $\widehat{C},\sigma$ or $\widehat{C} \pm k''\sigma$

Only the last is unambiguous, *provided that* the arbitrary multiplier k'' is specified. (In practice, one finds k's ranging from 1 to 6.) *Option 1*, per se, provides *no* information on the possible magnitude of the undetected component. *Option 2* is slightly ambiguous with respect to the tightness of the upper limit versus confidence level. For example, the limit $\leqslant 0.20$ μg/g (mercury in liver—Table 25.1) could have been obtained from -0.24 + 2 (0.22) or 0.080 + 2 (0.060). If k = 2 were replaced with k = 3, the first case would yield $<$ 0.42 μg/g, and the second, $<$ 0.26 μg/g. Including information on σ would of course eliminate the ambiguity. *Option 3* tells the user that the experimental result was indistinguishable from the blank, and it provides him with a fixed limit characteristic of the experimental procedure. The quoted limit should be the detection limit (*i.e.,* k' = 3.29), in order to avoid a discontinuity in upper limits between $\widehat{C}$'s which are just below C_C (*nd*) and just above (*d*).[22] *Option 4* is preferred, for it preserves all of the original information. It is the only option that permits a collection of results to be combined or properly intercompared. Nondetection may be indicated by *nd* or an asterisk.

COUNTING STATISTICS

Simple Counting

Application of the foregoing principles to counting simply requires the calculation of σ_0 from Poisson counting statistics. Taking the estimated concentration to be

$$\widehat{C} = \frac{N_p - N_b}{St} \tag{25.5}$$

Poisson statistics lead to

$$\sigma_0 = \sigma_{\widehat{C}} (C = 0) = \frac{\sqrt{\eta N_b}}{St} = \frac{\sqrt{\eta R_b/t}}{S} \tag{25.6}$$

where: N_p = gross peak counts
N_b = background counts
R_b = background count rate (counts/s)
t = counting time (seconds)
S = sensitivity function (counts/s per $\mu g/cm^2$)
η = 1 or 2 (cf. Equation 25.4)

Detection and quantitation limits are therefore

$$C_D = 3.29 \sigma_0 = 3.29 \sqrt{\eta N_b} / (St) \tag{25.7}$$

$$C_Q = 10 f \sigma_0 = 10 f \sqrt{\eta N_b} / (St) \tag{25.8}$$

The factor f corrects for the effects of small N_b (f $\approx$ 1 if $N_b > 2500$ counts). It follows from Equation 25.8 above, and Equation 15 in

Reference 9, that

$$f = \sqrt{1 + 25/(\eta N_b)} \ + \ \sqrt{25/(\eta N_b)} \qquad (25.9)$$

As an illustration, let us consider the analysis of potassium by energy-dispersive XRF. Using titanium as a secondary fluorescer with a spectrometer at the National Bureau of Standards, the following results were obtained:[17]

$$R_b = 2.83 \ (s^{-1}), \quad S = 17.2 \ (s^{-1} \ \mu g^{-1} \ cm^2)$$

Taking t = 100 s, $\eta = 1$ and assuming negligible uncertainties in the above parameters, we calculate

$$\begin{aligned} N_b &= 283 \text{ counts} \\ \sigma_0 &= 0.00978 \ \mu g/cm^2 \\ f &= 1.34 \end{aligned}$$

Thus from Equations 25.2 and 25.3 one obtains

$$\begin{aligned} C_D &= 0.032 \ \mu g/cm^2 \\ C_Q &= 0.131 \ \mu g/cm^2 \end{aligned}$$

For paired observations ($\eta = 2$) C_D and C_Q would be 0.046 $\mu g/cm^2$ and 0.17 $\mu g/cm^2$, respectively.

C_D and C_Q represent performance characteristics of the *measurement process*. If a particular *result* $(\widehat{C})$ were to exceed $C_C = 0.016 \ \mu g/cm^2$, one would conclude that potassium had been detected. The use of Equation 25.7 for C_D is illustrated by Jaklevic and Walter for the intercomparison of X-ray fluorescence spectrometers in Chapter 5 of this book. Another example of method performance, appraised according to the above criteria, has been given by McGinley and Schweikert.[13]   These authors have tabulated interference-free detection limits for X-rays from a large number of charged particle nuclear activation products.

Detailed treatment of more complex situations, such as fitting nonlinear baselines or complete spectra using least squares, is beyond the scope of this chapter. It should be noted, however, that relative standard errors (ϕ) from such analyses may be directly applied for judging procedure capabilities and results. Referring to Equation 25.1, for example, we see that any result whose relative standard deviation exceeds 61% ($\phi_C = 1/1.64$) may be considered *nd*. Similarly, $\phi_D \approx 30\%$ and $\phi_Q = 10\%$. Additional information on this topic may be found in References 8, 14 and 15.

Background Equivalent Concentration and Peak/Background

An alternative formulation in terms of the Background Equivalent Concentration (BEC) and the peak to background ratio r (reduced concentration) has two very important advantages: (1) it alerts one to examine background stability when r_D or r_Q becomes small, and (2) it provides the basis for a rapid graphical means for assessing detection and quantitation limits for any method, and it permits the simultaneous comparison of performance for several methods and/or elements as a function of counting time and analyte concentration.

If the background is interpreted as an equivalent concentration, then from Equation 25.5,

$$BEC = N_b/(St) \; = \; R_b/S \qquad\qquad (25.10)$$

BEC for the preceding example is therefore equal to $R_b/S = 2.83/17.2 = 0.165 \; \mu g/cm^2$. Similarly,

$$r_D = C_D/BEC \; = \; 3.29 \sqrt{\eta}/N_b \qquad\qquad (25.11)$$

and

$$r_Q = C_Q/BEC \; = \; 10 \; f \sqrt{\eta}/N_b \qquad\qquad (25.12)$$

where r_D and r_Q are the peak-to-background ratios at the detection and quantitation limits respectively. The importance of Equations 25.11 and 25.12 is that they depend *only* on the number of background counts. This permits the construction of a universal set of curves that applies regardless of sensitivity S or counting time t. Such curves are shown in Figure 25.2.

Once a counting time and analyte concentration have been selected for normalization, any procedure may be uniquely represented by a point in Figure 25.2. Concentration or time changes then correspond simply to y- or x-translations of the point, respectively.

Comparison of Alternative Methods

Four XRF methods for potassium are compared in Table 25.3, based upon background and sensitivity data provided by Pella[17] and Dzubay.[18] The comparison can be made more quickly and more flexibly, with respect to precision and counting time, by the graphical method of Figure 25.2. The quantities r_D and r_Q that were calculated with Equations 25.11 and 25.12 ($\eta = 2$), could have been deduced directly from the figure. For method 1 [λ], for example, the intersections of $N_b = 106$ counts

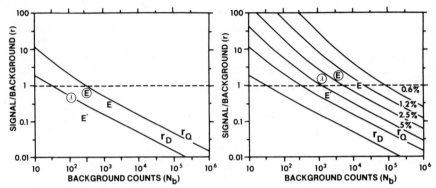

Figure 25.2 Method intercomparison, XRF analyses of potassium. Computed performance of the four methods of Table 25.3. Code: Ⓐ (method-1), Ⓔ (method-2), E (method-3), E' (method-4). The figure (2a) at the left is normalized to 0.10 $\mu g/cm^2$ and t = 100s; the figure at the right (2b) represents 0.30 $\mu g/cm^2$ and 10^3 s (Poisson statistics). Figure 25.2b includes added curves for estimating the RSD's between 0.6% and 5%.

Table 25.3 XRF Analysis of Potassium (K_α), Comparative Limits
(t = 100 sec, paired observations)

Dispersion[a]	$R_b(s^{-1})$	$S(\frac{cm^2}{s \cdot \mu g})$	$BEC(\frac{\mu g}{cm^2})$	r_D	$C_D(\frac{\mu g}{cm^2})$	$C_D'(\frac{\mu g}{cm^2})$[d]	$C_Q(\frac{\mu g}{cm^2})$
1. λ [b]	1.06	4.93	0.215	0.45	0.097	0.110	0.41
2. E (Ti)[b]	2.83	17.2	0.165	0.27	0.046	0.060	0.17
3. E (Ti)[c]	17.3	52.	0.333	0.112	0.037	0.086	0.123
4. E (Mo)[c]	1.3	2.1	0.62	0.41	0.25	0.29	1.04

[a]λ = wavelength dispersive, E = energy dispersive (fluorescer in parentheses).
[b]Data from P. Pella, National Bureau of Standards.[17]
[c]Data from T. Dzubay, Environmental Protection Agency.[18]
[d]Calculated from Equation 25.13, assuming 5% blank variability.

with the two curves yields directly r_D = 0.45 and r_Q = 1.9. Multiplication by BEC (Equations 25.11, 25.12) gives C_D and C_Q, i.e., $C_{D,Q} = r_{D,Q} \cdot BEC$.

Direct graphical intercomparison of all four methods is shown in Figure 25.2a, where the set of points has been normalized to 0.10 $\mu g/cm^2$ (K) and 100 s, respectively.* Confirming the tabular results, this figure

---

*Point Ⓔ (method 2), for example, lies at y = C/BEC = (0.10 $\mu g/cm^2$)/ (0.165 $\mu g/cm^2$) = 0.61, and x = $R_b \cdot t$ = (2.83 s^{-1})(100 sec) = 283.

indicates that method 1 [λ] lies close to its detection limit, and methods 2 [ⓔ] and 3 [E] near the quantitation limit. Figure 25.2b results from a translation of the set (relative positions of the points are invariant) by a factor of 10 horizontally (to 10^3 seconds) and a factor of 3 vertically (to 0.30 $\mu g/cm^2$). Under these circumstances, with the aid of the added RSD contours (ϕ = 0.6%, 1.2%, 2.5%, 5%), we see immediately that all methods have adequate precision for quantitative analysis, ranging from about 1.5% (method 3 [E]) to $\sim$ 8% (method 4 [E']).

THE REAL WORLD

Detection and quantitation limits as expressed in Equations 25.1-25.3 presumed that measurement errors were random, normal and estimable (σ_0). For Equations 25.6-25.9 it was further presumed that σ_0 could be deduced from the Poisson distribution. Such presumptions are never quite valid for real measurement processes or real samples. Nevertheless, the exercise is useful for two important reasons: (1) it presents a potentially attainable limit against which any measurement process should be compared (thus leading to possible improvement), and (2) it provides minimal standards (C_D, C_Q) for judging procedure adequacy.

Exceptions to the above assumptions occur with respect to control, imprecision and bias. Unless the overall procedure is in a state of control (stable with respect to random and systematic error) a Measurement Process cannot be said to exist. Certainly C_D and C_Q would be without meaning. Given control, if the magnitude of the systematic error can be sufficiently reduced, these performance characteristics can be usefully employed, even in the presence of non-Poisson imprecision.

Looking at Equation 25.5, one sees that non-Poisson errors in N_p, N_b and S are apt to be important. The accuracy of C_D in particular is extremely sensitive to interference and blank (N_b) effects. In this section we shall examine the more important error sources, as listed in Table 25.4.

Random Error

Imprecision may be estimated by replication of the *overall* measurement process. In practice this is seldom done, and therefore errors that could be random often become converted into systematic errors. If systematic error is negligible, one may use the estimated standard deviation s to compute meaningful confidence intervals, quantitation limits or detection limits. The use of Poisson statistics for computing the standard deviation,

Table 25.4   Assumption Limitations

(negligible bias + estimable imprecision $\rightarrow$ meaningful C_D, C_Q)

Random Error:	Poisson deviations from normality (N $\gtrsim$ 30)
	Random component of systematic error sources
	Random errors in corrections for systematic errors
Systematic Error:	Sampling and sample preparation (recovery)
	Blank, interference, and contamination
	Improper calibration and/or standards
	Matrix effects—particle size and composition, enhancement, absorption, and scattering
	Inaccurate data reduction models or correction formulas (assumed parameters, functional relations)
	Blunders and faulty reporting

when justified, is preferable, because it results in tighter (more precise) intervals and limits.

The Poisson assumption may be tested by comaring s^2 with σ^2; or, in least squares computations, by comparing χ^2/ν with its expected value (unity). This matter has been examined in some detail,[19] so it will not be repeated here. An important conclusion, however, is that excess (non-Poisson) variance is not trivial to detect: 10 replicates are required to detect random error whose standard deviation (σ_x) is twice that of the Poisson component (σ_p), and more than 450 are required if σ_x is half of σ_p!

Reproducibility of the blank is the key factor for the detection limit. Therefore, let us reexamine the example of the previous section (NBS measurements of potassium using energy-dispersive XRF and a titanium fluorescer), in the light of replicate measurements of N_b.[17]   Five observations of N_b for 500 s counting intervals were as follows:

$$N_b = 1413, 1449, 1391, 1436, 1394$$

The computed s and expected σ_p standard deviations are

$$s = 25.5 \qquad \sigma_p \approx \sqrt{N_b} = \sqrt{1417} = 37.6$$

The observed variability is consistent with that expected from counting statistics. With only four degrees of freedom, the 95% confidence interval for s/σ is 0.35 to 1.67.[19]   Therefore, at least for measurements *not exceeding 500 s*, the use of Poisson statistics is justified for this particular example.

The actual limit beyond which Poisson statistics may not be safely assumed for XRF depends very much upon the method, the laboratory and the sample. Scanning some of the recent literature suggests that the overall blank is seldom likely to be reproducible to much better than about 2% (RSD). The effect of such additional variability may be computed by incorporating the additional blank standard deviation term[19] ϵN_b, where ϵ = relative standard deviation, into Equation 25.7.

$$C'_D = \left(\frac{3.29 \sqrt{\eta}}{St} \right) \left(N_b + [\epsilon N_b]^2 \right)^{1/2} = C_D \left(1 + \epsilon^2 N_b \right)^{1/2} \qquad (25.13)$$

The importance of r_D, the peak to background ratio at the detection limit, for assessing sensitivity to background variability (or bias) is evident, when we apply Equation 25.13 to the data in Table 25.3. Method 3 was the most vulnerable of the four (r_D = 0.112); the 5% non-Poisson background variability (ϵ = 0.05) more than doubled its computed detection limit. It follows from Equations 25.11 and 25.13 that C_D will be seriously underestimated whenever $r_D \gtrsim 3.3 \sqrt{\eta} \; \epsilon$.

Systematic Error

Statistical confidence intervals and detection and quantitation limits are of little value in the presence of large systematic errors as indicated in Table 25.4. The most effective means for assessing accuracy is *via* intercomparisons—comparisons with standards or comparisons among laboratories or methods. One such exercise, carried out by Camp and co-workers, included both natural aerosol and synthetic (particles and evaporated solution) samples.[4] Measurements were reported by 22 investigators using 6 different analytical methods (including 3 XRF techniques). Especially pertinent was the finding that XRF (energy dispersive) was capable of analyzing synthetic samples with an accuracy of about 10% for elements above potassium. The overall bias for the particulate samples (micronized rock) was +5%, while that for the evaporated solution standards was only + 0.2%.

Reference 4 may be consulted also for information concerning the magnitudes of most of the errors indicated in Table 25.4. To cite one example, because of its significance for C_D, titanium contamination in the nylon reinforcing mesh used with certain filters yielded a blank value 1.5 times the average titanium signal from the deposit alone. The RSD of this blank was 2%. Blanks may arise not only from filter contamination (Dzubay and Stevens[3]), but also from subtle interference effects (Campbell *et al.*[5]). The latter authors noted that pulse pile-up of K- X-rays from 300 ppm of iron could produce a specious signal equivalent to 1 ppm of lead, a matter of some concern in the analysis of blood.

Systematic error may occur in the results of fitting multicomponent spectra due to omitted components or incorrect spectrum shapes.[14] Statistics cannot always protect the investigator from model-error bias, for significant errors may occur even when χ^2 is acceptable and when omitted components are not detectable.[15] This latter effect has been assessed quantitatively in Reference 16 for overlapping spectral peaks as a function of peak separation. There it was shown that the systematic error, due to the omission of the smaller peak from the model, exceeds the standard error if the peaks are closer than about one full-width-half-maximum, even when the smaller peak lies at its detection limit.

A more detailed discussion of error sources and magnitudes in XRF is outside the scope of this chapter. For those concerned with systematic errors and their correction, however, two additional references are Shenberg and Amiel[20] and Criss.[21]

Method Adequacy

The graphical approach used to compare the performance of the four alternative methods for potassium (Figure 25.2) is helpful also for the simultaneous comparison of a number of elements for a given method. A comparison of this sort is presented in Figure 25.3, based upon background and sensitivity data for one energy dispersive XRF system.[18]

Secondary fluorescers (for the nine elements considered) were: titanium (sulfur, potassium), molybdenum (iron, arsenic, selenium, strontium, lead) and samarium (cadmium, barium). The $L\beta$ X-ray was measured for lead; all others were K_α lines. Figure 25.3a represents the response of the spectrometer normalized to $C = 100$ ng/cm^2 for each element and $t = 100$ seconds. Coordinates for each of the points are computed as in Figure 25.2 ($y = C/BEC$, $x = R_b t$).

Figure 25.3a is the exact analog of Figure 25.2a; it shows us that, at the 100 ng/cm^2 level, a 100-second counting time is adequate to determine (RSD $\approx$ 10%) four of the elements (selenium, arsenic, strontium, potassium) and to detect four others (lead, iron, sulfur, cadmium); only barium remains well below its detection limit. Translation along the x-axis shows that $\sim$2000 seconds would be required to detect 100 ng/cm^2 of barium.

Figure 25.3a represents the ideal case of Poisson statistics. Figure 25.3b addresses the question of method adequacy for both actual

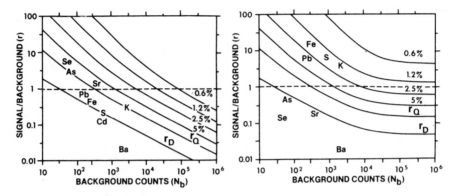

Figure 25.3 Method performance and adequacy. Computed performance of the energy-dispersive spectrometer[18] for nine trace elements. The figure at the left is normalized to 100 ng/cm² and 100s, for Poisson statistics; the figure at the right is normalized to "typical urban" concentrations and 100s and includes non-Poisson imprecision. Cadmium lies below the figure.

Vertical translations of the set of points reflect concentration changes; horizontal translations reflect changes in counting time. (See text for detailed explanation.)

concentrations of interest and in the presence of noncounting (random) error. The contours in Figure 25.3b differ from those in Figure 25.3a through the incorporation of additional variability (RSD) for background (1%) and for sensitivity (0.5%).[22] The y-coordinate of each elemental point has been adjusted to the "typical urban value" shown for the respective trace element in Chapter 5 of this book. For selenium the typical value was about 2 ng/cm²; thus, the y-coordinate of selenium in Figure 25.3b is decreased by a factor of 50 (100 ng/cm² ÷ 2 ng/cm²) as compared to Figure 25.3a.

Figure 25.3b is presented principally to stress the *qualitative* difference from the ideal Poisson situation (Figure 25.3a), when noncounting errors and actual concentrations of interest are considered. A striking difference in how the elements are grouped is evident. The elements potassium, iron, sulfur and lead are measurable with good precision (~ 2%) in 100s, arsenic and strontium are nearly detectable, and selenium, barium and cadmium are not detectable. The fine precisions for potassium, iron, sulfur and lead are a consequence of their high environmental concentrations. The asymptotic nature of the curves, which is due to background variability, shows that the current, low environmental concentrations of barium and cadmium will remain undetectable *regardless* of increases in counting time.

SUMMARY

Detection limits C_D and quantitation limits C_Q may be computed for X-ray fluorescence measurements given the assumption of normal, random errors, an estimate of the standard deviation of the blank σ_b and the sensitivity or calibration function S. Equations have been provided for calculating C_D and C_Q in the general case, when Poisson counting errors predominate and when other sources of error contribute. In all cases, the limits may be expressed in terms of RSD's of the net signal: 10% (quantitation), 30% (detection) and 61% (decision), *i.e.,* an observed result having an RSD $\geq$ 61% should be considered *nd* (not detected).

Different schemes for reporting data, particularly "not detected" results, continue to produce ambiguity and information loss. The preferred approach is always to report both the experimental result and its standard deviation, together with an indication of statistical significance (*d* or *nd*). Bounds for possible systematic error should also be included.

Comparison of XRF techniques is facilitated by a graphical approach in which reduced concentration (signal/background) is plotted versus background counts. This results in a single family of curves that are independent of time, sensitivity or background rate. Both numerical and graphical approaches have been illustrated for four XRF methods for the measurement of potassium in aerosol samples. Background equivalent concentration (BEC) and signal/background at the detection limit (r_D) have proved helpful in evaluating sensitivity to non-Poisson variations.

Factors such as sampling errors, chemical recovery, interference and contamination, matrix absorption and scattering and data reduction models all pose limitations to the above assumptions. Assessment of imprecision via replication, and assessment of systematic errors via intercomparison yields information concerning the range over which the Poisson assumption may be safely applied. However, conclusions concerning the adequacy of fit of regression models, especially in the case of overlapping peaks, should be accepted with some caution.

Finally, the detection and measurement precision of nine trace elements, for a particular energy dispersive XRF spectrometer, were evaluated graphically in two modes. First, a fixed concentration was taken for all of the elements and Poisson statistics were assumed. Second, actual measurement adequacy as a function of counting time was treated by normalizing to typical urban concentrations and introducing some allowance for non-Poisson imprecision. A dramatic rearrangement occurred in the second case, showing (for these particular levels and assumptions) that potassium, iron, sulfur and lead could be determined with excellent precision with 100-second observations, whereas even extended counting would not permit barium or cadmium to be reliably detected.

ACKNOWLEDGMENTS

Thanks are due to T. Dzubay and to P. Pella for illustrative data from their XRF spectrometers as well as for much helpful discussion. Editorial assistance from R. Murphy is gratefully acknowledged. Appreciation goes also to K. Heinrich for introducing the author to the field, its instrumentation and its problems.

REFERENCES

1. Goldberg, E. D. "Baseline Studies of Heavy Metal, Halogenated Hydrocarbon, and Petroleum Hydrocarbon Pollutants in the Marine Environment and Research Recommendations," *Deliberations of the International Decade of Ocean Exploration (IDOE) Baseline Conference* (1972).
2. EPA-NBS Interlaboratory Comparison for Chemical Elements in Coal, Fly Ash, Fuel Oil and Gasoline (1973).
3. Dzubay, T. G. and R. K. Stevens. "Application of X-Ray Fluorescence to Particulate Measurements," *Second Joint Conference on Sensing of Environmental Pollutants* (1973).
4. Camp, D. C., A. L. VanLehn, J. R. Rhodes and A. H. Pradzynski. "Intercomparison of Trace Element Determinations in Simulated and Real Air Particulate Samples," *X-Ray Spectrometry* 4:123-137 (1975).
5. Campbell, J. L., B. H. Orr, A. W. Herman, L. A. McNelles, J. A. Thomson, and W. B. Cook. "Trace Element Analysis of Fluids by Proton-Induced X-Ray Fluorescence Spectrometry," *Anal. Chem.* 47:1542 (1975).
6. Heinrich, K. F. S. "Common Sources of Error in Electron Probe Microanalysis," *Advances in X-Ray Analysis* 11:40 (1968).
7. Heinrich, K. F. S. "Empirical Approaches for the Treatment of Interelement Effects," to be published in *Advances in X-Ray Analysis* (1976).
8. Müller, R. O. *Spectrochemical Analysis by X-Ray Fluorescence* (New York: Plenum Press, 1972), See especially Chapter 11 and Part III.
9. Currie, L. A. "Limits for Qualitative Detection and Quantitative Determination. Application to Radiochemistry," *Anal. Chem.* 40: 586 (1968).
10. Dixon, W. J. and F. J. Massey, Jr. *Introduction to Statistical Analysis* (New York: McGraw-Hill, 1957).
11. Bevington, P. R. *Data Reduction and Error Analysis for the Physical Sciences* (New York: McGraw-Hill, 1969).
12. Eisenhart, C. "Expression of the Uncertainties of Final Results," *Science* 160:1201 (1968).
13. McGinley, J. R. and Schweikert, E. A. "Multielement Charged Particle Activation Analysis with X-Ray Counting," *Anal. Chem.* 48:429 (1976).

14.   Gehrke, R. J. and R. C. Davies. "Spectrum Fitting Technique for Energy Dispersive X-Ray Analysis of Oxides and Silicates with Electron Microbeam Excitation," *Anal. Chem.* 47:1537 (1975).

15.   Currie, L. A. "The Discovery of Errors in the Detection of Trace Components in Gamma Spectral Analysis," *Modern Trends in Activation Analysis,* Vol. II, J. R. DeVoe and P. D. LaFleur, Ed., 1215-30, U.S. National Bureau of Standards Special Publication 312 (1968).

16.   Currie, L. A. "Evaluation of Nuclear Spectra: Incomplete Models and Systematic Error," *Modern Trends in Activation Analysis Conference,* Munich (1976).

17.   Background and sensitivity data for potassium using energy- and wavelength-dispersive XRF spectrometers at the NBS were supplied by P. A. Pella.

18.   T. G. Dzubay provided background and sensitivity data for an older EPA energy-dispersive XRF spectrometer. Backgrounds quoted were obtained with a 5 mg/cm^2 cellulose ester substrate.

19.   Currie, L. A. "The Limit of Precision in Nuclear and Analytical Chemistry," *Nucl. Instr. Meth.* 100:387-395 (1972).

20.   Shenberg, C. and S. Amiel. "Critical Evaluation of Correction Methods for Interelement Effects in X-Ray Fluorescence Analysis Applied to Binary Mixtures," *Anal. Chem.* 46:1512 (1974).

21.   Criss, J. W. "Particle Size and Composition Effects in X-Ray Fluorescence Analysis of Pollution Samples," *Anal. Chem.* 48:179 (1976).

22.   Currie, L. A. "The Evaluation of Radiocarbon Measurements and Inherent Statistical Limitations in Age Resolution," in *Proc. Eighth International Conference on Radiocarbon Dating* (Royal Society of New Zealand, 1973).